Frederic P. Miller, Agnes F. Vandome,
John McBrewster (Ed.)

Dodge Charger (B-body)

Frederic P. Miller, Agnes F. Vandome,
John McBrewster (Ed.)

Dodge Charger (B-body)

Dodge Charger, Car model, Automobile, Dodge, Chrysler B platform, Chrysler Cordoba, Chrysler, Muscle car, List of automobile model nameplates with a discontiguous timeline

Alphascript Publishing

Imprint

Permission is granted to copy, distribute and/or modify this document under the terms of the GNU Free Documentation License, Version 1.2 or any later version published by the Free Software Foundation; with no Invariant Sections, with the Front-Cover Texts, and with the Back-Cover Texts. A copy of the license is included in the section entitled "GNU Free Documentation License".

All parts of this book are extracted from Wikipedia, the free encyclopedia (www.wikipedia.org).

You can get detailed informations about the authors of this collection of articles at the end of this book. The editors (Ed.) of this book are no authors. They have not modified or extended the original texts.

Pictures published in this book can be under different licences than the GNU Free Documentation License. You can get detailed informations about the authors and licences of pictures at the end of this book.

The content of this book was generated collaboratively by volunteers. Please be advised that nothing found here has necessarily been reviewed by people with the expertise required to provide you with complete, accurate or reliable information. Some information in this book maybe misleading or wrong. The Publisher does not guarantee the validity of the information found here. If you need specific advice (f.e. in fields of medical, legal, financial, or risk management questions) please contact a professional who is licensed or knowledgeable in that area.

Any brand names and product names mentioned in this book are subject to trademark, brand or patent protection and are trademarks or registered trademarks of their respective holders. The use of brand names, product names, common names, trade names, product descriptions etc. even without a particular marking in this works is in no way to be construed to mean that such names may be regarded as unrestricted in respect of trademark and brand protection legislation and could thus be used by anyone.

Cover image: www.PureStockX.com
Concerning the licence of the cover image please contact PureStockX.

Publisher:
Alphascript Publishing is a trademark of
VDM Publishing House Ltd.,17 Rue Meldrum, Beau Bassin,1713-01 Mauritius
Email: info@vdm-publishing-house.com
Website: www.vdm-publishing-house.com

Published in 2010

Printed in: U.S.A., U.K., Germany. This book was not produced in Mauritius.

ISBN: 978-613-0-29334-5

Contents

Articles

Dodge Charger (B-body)	1
Dodge Charger	11
Car model	12
Automobile	14
Dodge	28
Chrysler B platform	40
Chrysler Cordoba	42
Chrysler	46
Muscle car	53
List of automobile model nameplates with a discontiguous timeline	68

References

Article Sources and Contributors	70
Image Sources, Licenses and Contributors	73

Dodge Charger (B-body)

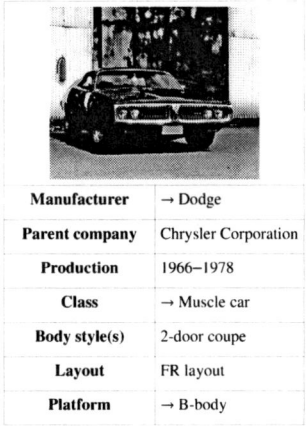

Manufacturer	→ Dodge
Parent company	Chrysler Corporation
Production	1966–1978
Class	→ Muscle car
Body style(s)	2-door coupe
Layout	FR layout
Platform	→ B-body

The **Dodge Charger** was a → model of car produced by → Dodge. The 1966 to 1974 Chargers were sporty models based on the → Chrysler B platform that could be ordered with high-performance options. The 1975 to 1978 Chargers were based on the → Chrysler Cordoba. The Dodge Charger R/T was one of the largest muscle cars available in the 1970s.

Birth of the Charger

In 1964, when the Pontiac GTO started the American → muscle car era with strong sales, the rest of GM's divisions were quick to jump on the muscle car bandwagon. Buick followed with the Gran Sport and even Oldsmobile brought out the 442. Dodge, despite putting out cars that could meet or beat these cars on the street or strip, didn't have a performance image muscle car of their own. Even with available performance engines, the Coronet's styling and image was considered by most to be "conservative."

1965 Dodge Charger II Show Car

Burt Bouwkamp, Chief Engineer for Dodge during the 1960s and one of the men behind the → Dodge Charger, related his experience during a speech in July 2004.

"Lynn Townsend was at odds with the Dodge Dealers and wanted to do something to please them. So in 1965 he asked me to come to his office - for the second time. He noted that one of the Dodge Dealer Council requests was for a Barracuda type vehicle. The overall dealer product recommendation theme was the same - we want what Plymouth has. The specific request for a Mustang type vehicle was not as controversial to Lynn. His direction to me was to give them a specialty car but he said 'for God's sake don't make it a derivative of the Barracuda': i.e. don't make it a Barracuda competitor.

"So the 1966 Charger was born.

"We built a Charger 'idea' car which we displayed at auto shows in 1965 to stimulate market interest in the concept. It was the approved design but we told the press and auto show attendees that it was just an "idea" and that we would build it if they liked it. It was pre-ordained that they would like it."[1]

The concept car received a positive response, so Dodge put it into production.

1966-1967

1966

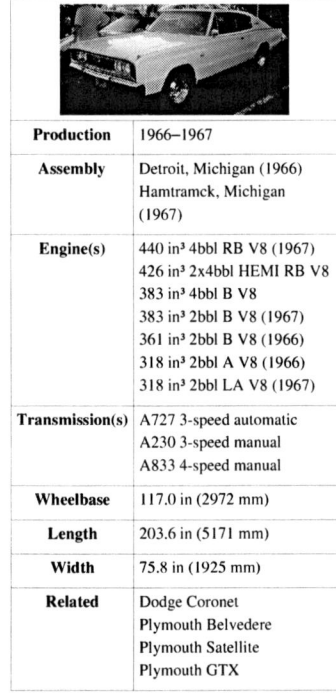

Production	1966–1967
Assembly	Detroit, Michigan (1966) Hamtramck, Michigan (1967)
Engine(s)	440 in³ 4bbl RB V8 (1967) 426 in³ 2x4bbl HEMI RB V8 383 in³ 4bbl B V8 383 in³ 2bbl B V8 (1967) 361 in³ 2bbl B V8 (1966) 318 in³ 2bbl A V8 (1966) 318 in³ 2bbl LA V8 (1967)
Transmission(s)	A727 3-speed automatic A230 3-speed manual A833 4-speed manual
Wheelbase	117.0 in (2972 mm)
Length	203.6 in (5171 mm)
Width	75.8 in (1925 mm)
Related	Dodge Coronet Plymouth Belvedere Plymouth Satellite Plymouth GTX

Carl "CAM'" Cameron would be the exterior designer of Dodge's new flagship vehicle, and on January 1, 1966, viewers of the Rose Bowl were first introduced to the new "Leader of the Dodge Rebellion", the 1966 Charger. The Charger's debut followed by a half model year Chrysler corporation's tremendously successful introduction of a new street version of the 426 Hemi (7.0 L) in its Plymouth line. Finally, Dodge would have a performance platform, and a performance image, to go along with this high performance engine.

As the 1966 Charger's features would go, the "electric shaver" grille used fully rotating headlights, not seen on a Chrysler product since the 1942 DeSoto, that when opened or closed made the grille look like one-piece. Inside, the Charger used four individual bucket seats with a full length console from front to rear. The rear seats and console pad also folded forward, and the trunk divider dropped back, which allowed for lots of cargo room inside. Many other things were exclusive to the Charger such as the door panels, courtesy lights and the instrument panel.

1966 Charger dash

The instrument panel was especially interesting as regular bulbs weren't used to light the gauges. Instead four electroluminescent dash pods housed the tachometer, speedometer, alternator, fuel and temperature gauges. In the rear the full length taillight read CHARGER.

The engine selection was all V8s. A six cylinder engine didn't make the option list until 1968. In 1966 four engines were offered; the base-model 318 in³ (5.2 L) 2-barrel V8, the truck-sourced 361 in³ (5.9 L) 2-barrel, the 383 in³ (6.3 L) 4-barrel, and the new 426 Street Hemi. The majority of 1966 Chargers were ordered with the 325 hp (242 kW) 383.

Total production in 1966 came to 37,344 units, which was successful for the mid-year introduction.

In 1966 Dodge took the Charger into NASCAR in hopes that the fastback would make their car a winner on the high-banks. But the car proved to have rear end lift around corners which made it very slippery on the faster tracks. The lift was because the air actually traveled faster over the top of the car than under it, causing the car to act like a giant airplane wing. Drivers would later claim that "it was like driving on ice." In order to solve this problem Dodge installed a small lip spoiler on the trunk lid which improved traction at speeds above 150 mph (240 km/h). They also had to make it a dealer-installed option in late 1966 and through 1967 because of NASCAR rules (with small quarter panel extensions in 1967). The 1966 Charger was the first US production vehicle to offer a spoiler. David Pearson, driving a #6 Cotten Owens-prepared Charger, went on to win the NASCAR Grand National championship in 1966 with 14 first-place finishes.

1967 Charger NASCAR Spoiler

1967

Since the Charger was such a sales success despite its midyear introduction, changes were limited for 1967. Outside, new fender-mounted turn signals were introduced and would serve as the main outside identifier between a 1966 and 1967 Charger. A vinyl roof become available as well. Inside, the full length console was gone, due in part to customer complaints about entry and exit from the back seats. It was replaced with a regular sized console. Bucket seats were standard, but a folding armrest/seat and column shifter was an option allowing three people to sit up front.

As for engine options, the 440 "Magnum" was added and the 361 in³ engine was replaced by a 383 in³ engine. The 440 was conservatively rated at 375 hp (280 kW) with a single 4-barrel carburetor. The 318 two-barrel engine remained, although it was now the modern Chrysler LA engine with wedge-shaped combustion chambers, unlike the outdated 1966 polyspherical (or "poly") design. The 383 4-barrel and the 426 Street Hemi remained as options.

Despite the Chargers' NASCAR racing success of 1966, sales slipped by half. In 1967 only 15,788 Chargers were sold. The Chargers faced competition from the Trans-Am Series, the Ford Mustang and the just introduced Chevrolet Camaro. Dodge decided that a major redesign was in order, rather than a minor face-lift.

1968-1970

1968

Production	1968–1970
Assembly	Detroit, Michigan
	Hamtramck, Michigan
	Los Angeles, California
	St. Louis, Missouri
Engine(s)	225 in³ 1bbl Slant Six A (1969-70)
	318 in³ 2bbl LA V8
	383 in³ 2bbl B
	383 in³ 4bbl B
	426 in³ HEMI 2x4bbl RB V8
	440 in³ 4bbl RB V8
	440 in³ 2x3 RB (1970)
Transmission(s)	A904 3-Speed Automatic
	A727 3-speed automatic
	A230 3-speed manual
	A833 4-speed manual
Wheelbase	117.0 in (2972 mm)
Length	208.0 in (5283 mm)(1968) 207.9 in (5281 mm) (1969-70)
Width	76.7 in (1948 mm) (1968-69) 76.6 in (1946 mm) (1970)
Height	53.2 in (1351 mm) (1968-69) 53.0 in (1346 mm) (1970)
Related	Dodge Coronet
	Plymouth Belvedere
	Plymouth Satellite
	Plymouth GTX
	Plymouth Road Runner

It was clear after the sales drop of the 1967 Charger that a restyle was in order. Dodge was going to restyle their entire B-body lineup for 1968 and decided that it was time to separate the Coronet and Charger models even further. What designer Richard Sias came up with was a double-diamond design that would later be referred to as coke bottle styling. From the side profile the curves around the front fenders and rear quarter panels look almost like a Coke bottle. Front and rear end sheet metal was designed by Harvey J Winn. The rear end featured a "kick up" spoiler appearance, inspired by Group 7 racing vehicles. On the roof a "flying buttress" was added to give the rear window area a look similar to that of the 1966-67 Pontiac GTO. The Charger retained its full-length hidden headlight grille, but the fully rotating electric headlights had been replaced by a simple vacuum operated cover, similar to the Camaro RS. The full length taillights were gone as well. Instead, dual Corvette-inspired taillights were added at the direction of Styling Vice President, Elwood P. Engel. Dual scallops were added to the doors and hood to help accent the new swoopy lines. Inside, the interior shared almost nothing with its first generation brothers. The four bucket seats were gone, the console remained the same as the '67. The tachometer was now optional instead of standard, the trunk and grille medallions were gone, the carpeting in the trunk area was gone, replaced by a vinyl mat, the rear seats did not fold forward and the space-age looking electroluminescent gauges disappeared in favor of a more conventional looking design.

In order to further boost the Charger's muscle car image, a new high-performance package was added, the R/T. This stood for "Road/Track" (no 'and' between Road and Track) and would be the high performance badge that would establish Dodge's performance image. Only the high performance cars were allowed to use the R/T badge. The R/T came standard with the previous year's 440 "Magnum" and the 426 Hemi was optional. The standard engine was the 318 2bbl the rest of the engine lineup (383-2, 383-4) remained unchanged.

In 1968 Chrysler Corporation unveiled a new ad campaign featuring a Bee with an engine on its back. These cars were called the "Scat Pack". The Coronet R/T, Super Bee, Dart GTS and Charger R/T received bumble-bee stripes (two thin stripes framing two thick stripes). The stripes were standard on the R/Ts and came in red, white or black. They also could be deleted at no cost. These changes and the new Charger bodystyle proved to be very popular with the public and helped to sell 96,100 Chargers, including over 17,000 Charger R/Ts.

A famous Charger was the four-speed, triple-black 1968 Charger R/T used in the movie *Bullitt*. The chase scene between Steve McQueen's fastback Mustang GT and the hitmen's Charger R/T is popularly regarded as one of the greatest car chase scenes ever filmed. During filming of the scene, the Charger proved to be extremely durable. When performing the various jumps over the hills in San Francisco, the Mustang GT encountered several suspension problems, while the suspension of the Chargers used never failed once.

1969

1969 Dodge Charger

In 1969 not much was changed for the popular Charger. Exterior changes included a new grille with a center divider and new longitudinal taillights both designed by Harvey J. Winn. A new trim line called the Special Edition (SE) was added. This could be available by itself or packaged with the R/T, thus making an R/T-SE. The SE added leather inserts to the front seats only, chrome rocker moldings, a wood grain steering wheel and wood grain inserts on the instrument panel. A sunroof was added to the option list as well, and it would prove to be a very rare option (some 260 sold). The bumble bee stripes returned as well, but were changed slightly. Instead of four stripes it now featured one huge stripe framed by two smaller stripes. In the middle of the stripe an R/T cutout was placed. If the stripe was deleted, then a metal R/T emblem was placed where the R/T cutout was. Total production was around 89,199 units. But in 1969 Dodge had its eye on NASCAR and in order to compete it would have to create two of the most rare and desirable of all Chargers: ***Charger 500***, and the ***Charger Daytona***.

The television series *The Dukes of Hazzard* (1979-1985) featured a 1969 Dodge Charger that was named *The General Lee*, often noted as being the most recognizable car in the world. "The General" sported the Confederate battle flag painted on the roof and the words "GENERAL LEE" over each door. The windows were always open, as the doors were welded shut. The number "01" is painted on both doors. Also, when the horn button was pressed, it played the first 12 notes from the *de facto* Confederate States anthem "Dixie". The muscle car performed spectacular jumps in almost every episode, and the show's popularity produced a surge of interest in the car. The show itself purchased hundreds of Chargers for stunts, as they generally destroyed at least one car per episode. (Real

Bo and Luke Duke popularized the 1969 Dodge Charger in *The Dukes of Hazzard*

Chargers stopped being used for jumps at the end of the show's sixth season, and were begrudgingly replaced with miniatures.) The Dodge Charger in "The Fast And The Furious" was a 1969 Charger remodelled to look like a 1970 Charger

Charger 500

Dodge Charger 500

In 1968, Dodge watched their NASCAR inspired Charger R/T fail to beat the Ford cars on the high-banks oval-tracks. The Dodge engineers went back to the wind tunnel and found the tunneled rear window caused lift and the gaping mouth induced drag. Dodge engineers made the rear window flush with the rest of the hood and put a 1968 Coronet Grille up front. The original Charger 500 prototype was a 1968 Charger R/T with a 426 Hemi. The prototype was painted in B5 Blue with a white stripe.

The Charger 500 prototype had a Torqueflite, a white interior and 426 Hemi. The Charger 500 was tested for production, got the greenlight and was one of three models introduced in September 1968. The Charger 500 was standard with the 440 Magnum but the factory literature claims the 426 Hemi was standard. The Charger 500 had the Torqueflite standard and the same equipment standard as the R/T.

The Charger 500 sold for $3842.00 MSRP. The Charger 500 was available with the 426 Hemi for an extra $648.20, The Charger 500 had the options A11 and V88 the stripe was available with red, black and white colors. The Charger 500 was available with Air Conditioning on the 440 Magnum for an extra $357.65.

A total of 500 Charger 500s were made but only 392 were bought for street use. The rest were bought by racers and gutted, stripped, modified and/or repainted. Only 67 Charger 500s were built with the 426 Hemi; 27 with 4-Speeds and 40 with Torqueflites.

The Charger 500 did not get the results expected of it on the NASCAR circuits and lost to FoMoCo entrants.

Charger Daytona

Dodge was not satisfied with the results of the Charger 500. The car was not enough to beat the other aerocars on the NASCAR circuit. After months of research and development, including the aftermarket shop Creative Industries Inc., the Dodge Charger Daytona was introduced on April 13, 1969. Within hours of its unveiling, Dodge had received over 1,000 orders, despite the price point of $3,993.00 MSRP.

Chrysler made many attempts at improving the aerodynamics of the 500 by adding noses rumoured to be up to 23 inches (580 mm) long. The Charger Daytona finally received an 18-inch (460 mm) nose. The full size Charger Daytona was tested with an 18-inch (460 mm) nose at the Lockheed-Martin Georgia facility. The test was a success and the project was greenlighted. The nose piece was only part of the innovation. The Charger Daytona also received a 23-inch (580 mm) tall wing in rear. This wing was bolted through the rear quarter panels and into the rear subframe. The Charger Daytona's wing also helped out in an unintended way, by giving the car directional stability as well.

The Charger Daytona proved itself to high management but was shot down by Dodge's styling department. The Charger Daytona engineering model was tested on the Chelsea, Michigan Chrysler Proving Grounds on July 20, 1969. Driven by Charlie Glotzbach and Buddy Baker, it was clocked at 205 mph (330 km/h) with a small 4 bbl. carb. The Charger Daytona's nose made 1,200 pounds of downforce and the wing made 600 pounds of downforce. (a zero lift car) The Dodge styling department wanted to make changes to the Charger Daytona as soon as they saw it, but was told by Bob McCurry to back off; he wanted function over finesse.

The Charger Daytona introduced to the public had a fiberglass nose without real headlamps and a wing without streamlined fairings. The media and public loved the car, but were mystified by the reverse scoops on the front fenders. The PR representatives claimed it was for tire clearance. Actually, they reduced drag 3%.

The Charger Daytona came standard with the 440 Magnum Engine with 375 hp (280 kW) and 480lb.-ft. of torque, A727 Torqueflite Automatic Transmission, and a 3.23 489 Case 8 3/4 Chrysler Differential. The Charger Daytona also came with the 426 Hemi with 425 hp (317 kW) and 490 lb·ft (660 N·m) (620 hp (460 kW) at 6000 rpm and 620 lb·ft (840 N·m) at 4700 rpm) for an extra $648.20. The 426 Hemi was also available with the no cost option of the A833 4-Speed Manual. Only 503 Charger Daytonas were built, 433 were 440 Magnum 139 4-Speed and 294 Torqueflite; 70 were 426 Hemi power, 22 4-Speed and 48 Torqueflite.

In the end the Daytona was brought down by the decision to make the 1970 Plymouth Superbird the only 1970 aerocar, however apparently two Charger Daytonas were built using 1970 sheet metal. One of them resides in the backyard of an individual in western Pennsylvania, not far from Pittsburgh. Several enthusiasts have tried to buy the car, but the owner refuses to sell, and the car sits un-covered and deteriorating. While Daytona's were raced through the 1970 season, only one Daytona still raced until 1971 (in the 1971 Daytona 500) when NASCAR decreed that engine displacement of wing cars would be limited to 305 ci. That particular car finished somewhere in the top-10 of that race.

1970

In 1970 the Charger changed slightly again. This would be the last and rarest year of the 2nd generation Charger and it now featured a large wraparound chrome bumper and the grille was no longer divided in the middle. New electric headlight doors replaced the old vacuum style. Side markers were now actual lights. The taillights were similar to those used in 69, but 500 and R/T models came with a new more attractive taillight panel. On the R/T new rear-facing scoops with the R/T logo were mounted on the front doors, over the door scallops. A new 440 or HEMI hood cutout made the option list for this year only.

In order to achieve the desired look, Dodge painted the hood scallop inserts black and put the silver engine callouts on top. New "High Impact" colors were given names, such as Top Banana, Panther Pink [2], Sublime, Burnt Orange, Go Mango and Plum Crazy (sometimes nicknamed "Statutory Grape"[3]). The 500 returned for another year, but now it was just a regular production Charger unlike the limited production NASCAR Charger of 1969.

Interior changes included new high-back bucket seats, the door panels were also revised and the map pockets were now optional instead of standard. The ignition was moved from the dash to the steering column (as with all Chrysler products this year), and the glove box was now hinged at the bottom instead of the top as in 1968-69. The SE "Special Edition" option added high end luxury to a full on muscle car and was available as 500 SE and R/T SE models. The all new pistol grip shifter was introduced, along with a bench seat, a first for the Charger since its debut.

A new engine option made the Charger's list for the first time, the 440 Six Pack. With three two-barrel carburetors and a rating of 390 hp (291 kW), it was one of the most exotic setups since the cross-ram Max Wedge engines of the early 1960s. The Six Pack was previously used on the mid-year 1969 Dodge Super Bee and Plymouth Road Runner and was notorious for beating the Hemi on the street. Despite this hot new engine, production slipped again to 46,576 but most of this was due to the brand new E-body Dodge Challenger and the high insurance rates. In the 1970 Nascar season it was the 1970 Charger that tallied up more wins (10) than any other car....including the notorious 69 Dodge Charger Daytonas and Plymouth Superbirds, giving Bobby Isaac the Grand National Championship.

Dodge Charger R/T

Dodge Charger

1968 Dodge Charger R/T

1968 Dodge Charger R/T

1971-1974

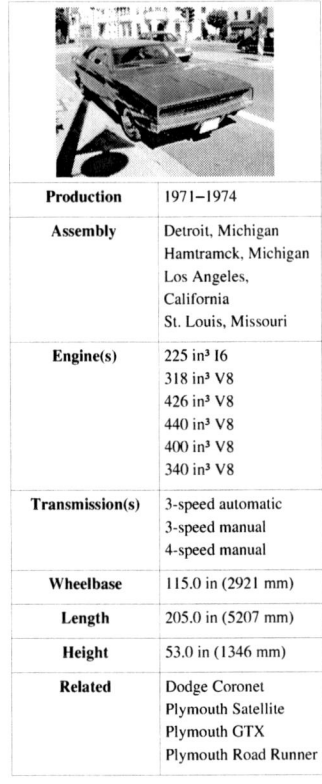

Production	1971–1974
Assembly	Detroit, Michigan Hamtramck, Michigan Los Angeles, California St. Louis, Missouri
Engine(s)	225 in³ I6 318 in³ V8 426 in³ V8 440 in³ V8 400 in³ V8 340 in³ V8
Transmission(s)	3-speed automatic 3-speed manual 4-speed manual
Wheelbase	115.0 in (2921 mm)
Length	205.0 in (5207 mm)
Height	53.0 in (1346 mm)
Related	Dodge Coronet Plymouth Satellite Plymouth GTX Plymouth Road Runner

In 1971, the all-new third generation Charger debuted. It was completely restyled with a new split grille and more rounded "fuselage" bodystyle. The interiors now looked more like those of the E-body and were now shared by the Plymouth B-body. Hidden headlights were no longer standard, they were now optional. A rear spoiler and a "Ramcharger" hood made the option lists for the first time. A special scoop was mounted in the hood, directly above the air cleaner. If the driver wanted to draw clean air directly into the carburetor, he pulled a small lever under the dash and the scoop popped up. The Plymouth Roadrunner used this device and called it the "air grabber hood". This device had been used on the Coronet R/T and Super Bees, but this was the first time it was used on the Charger.

Dodge also merged its Coronet and Charger lines. From 1971, all four-door B-bodies were badged as Coronets and all two-door B-bodies as Chargers. This change would add the one-year-only Charger Super Bee to the Charger stable.

The Dodge Super Bee made the move from the Coronet line to the Charger line for 1971 only, then the model was discontinued. Several other models were carried over from 1970, including the 500. The R/T and SE versions carried over as well, but the R/T's popularity was on the downslide thanks to higher insurance costs. Only 63 Hemi versions were built, and 2,659 were built with other engines that year. Rapidly rising insurance rates, combined with higher gasoline prices, reduced sales of muscle cars and 1971 was the last year of availability for the 426 Hemi "Elephant engine" in any car. 1971 also saw the end of the high-performance 440 Six-Pack engine (although some Dodge literature stated that this engine was available for 1972, it was pulled at the last minute. However, a few factory installed six-pack Chargers and Road Runners were built very early in the production run).

The 1972 Charger bowed with a new "Rallye" option to replace the former R/T version. The SE was differentiated from other 1972 Chargers by a unique formal roof treatment and hidden headlights. The 440 engines were still available, but now had to use the net horsepower rating instead of the gross horsepower rating. This would cause their horsepower ratings to go down substantially, although the net horsepower rating was actually more realistic. Also beginning in 1972, all engines featured lowered compression ratios to permit the use of regular leaded or unleaded gasoline rather than leaded premium fuel as in past years due to increasing tighter emissions regulations. A low-compression 440 with a 4 barrel carburetor became the top dog engine, and the use of the pistol-grip 4-speed Hurst manual shifter was limited to engines of 400 cubic inches.

Unusual triple opera window on 1973 Dodge Charger SE

The 1973 Chargers sported new vertically slatted taillights and new grilles (and no more hidden headlights, even as an option). The 318 was still standard, with the 340 (available only on the Rallye), 400 and 440 remaining as options. The SE models had a new roof treatment that had "triple opera window" treatment surrounded by a canopy-style vinyl roof. All other models had a new quarter window treatment, ditching its AMC Gremlin-style window in favor of a more conventional design. Sales this year were around 108,000 units, the highest ever for the 1971-74 Charger generation.

1974 was a virtual rerun of 1973. Minor changes included all new color choices, a softer grain pattern on interior surfaces, and a slight increase in the size of the rubber bumper tips(brought on by ever-changing federal front and rear impact regulations). The biggest news was that the 340 option was dropped and the 360 4bbl replaced the 340 as the small block performance engine. All other engine options remained the same. Several performance rear end ratios, including a 3.23 limited slip rear end were still available. A four speed transmission was still an option except with the 440 engine. Emphasis in these years turned to luxury instead of performance, hence the high sales figures for the SE model, but one could still equip a Charger with respectable performance options if one were so inclined and turn in decent performance figures for the day. The Charger, however, was no longer considered a performance car, and was gradually turned into personal luxury car, because all manufacturers "saw the handwriting on the wall." The muscle car era came to a close, and the 1975 Dodge Charger would be the final nail in the coffin.

The 1971-74 Chargers were campaigned in NASCAR, with Buddy Baker, Bobby Issac, Dave Marcis, and Richard Petty scoring several wins. *Petty* won 25 races with this body style bewtween 1972 and 1977 (NASCAR allowed the Chargers to run a few years longer than normal, as Chrysler had did not have anything else to replace it) and it is Petty's self proclaimed favorite car that he ran in his career.

1975-1978

Dodge Charger (B-body)

Production	1975–1978
Assembly	Detroit, Michigan Hamtramck, Michigan Los Angeles, California St. Louis, Missouri
Engine(s)	318 in³ V8 360 in³ V8 400 in³ V8
Transmission(s)	3-speed automatic 3-speed manual 4-speed manual
Wheelbase	115.0 in (2921 mm)
Length	216.0 in (5486 mm)
Height	52.0 in (1321 mm)
Related	Dodge Coronet → Chrysler Cordoba Plymouth GTX Plymouth Fury

Beginning in 1975, the Dodge Charger was based on the Chrysler Cordoba. The Charger SE (Special Edition) was the only model offered. It came with a wide variety engines from the 318 in³ (5.2 L) "LA" series small block V8 to the 400 in³ (6.6 L) big block V8. The standard engine was the 360 in³ (5.9 L) small block. Sales in 1975 amounted to 30,812. Because of the extreme squareness of the bodystyle, NASCAR teams were forced to rely on the previous years (1974) sheetmetal for race-spec cars. In order for Dodge to be represented, NASCAR allowed the 1974 sheetmetal to be used until January 1978, when the new Dodge Magnum was ready for race use.

In 1976 the model range was expanded to four models — base, Charger Sport, Charger SE and the Charger Daytona. The base and Sport models used a different body than the SE and Daytona, and were essentially a rebadging of what had been the 1975 Dodge Coronet 2-door models - and available with a 225 in (3.7 L) Slant Six, which was not offered on the SE and Daytona. The Charger Daytona was introduced in hopes or rekindling the performance fire, but it amounted to little more than a tape/stripe package. It did offer either the 360 small block or the 400 big block. Sales did go up slightly to 65,900 in 1976 but would quickly plummet after that mainly due to the fact the base and Sport models were one-year only offerings that did not return for 1977.

In 1977 the base Charger and Charger Sport were dropped as this body style became part of the newly named B-body Monaco line, and only the Charger SE and Charger Daytona were offered. Sales dropped to 36,204. In 1978 only about 2,800 Chargers were produced (likely to use up leftover stock of 1977 trim parts), after which it was replaced by the similar 1978 Dodge Magnum.

See also

- → List of automobile model nameplates with a discontiguous timeline

External links

- Dodge Charger [4] at the Open Directory Project

References

[1] http://www.allpar.com/cars/dodge/charger-history.html
[2] http://www.pantherpink.com
[3] Four Wheel Drift: Best and Worst Paint Colors (http://fourwheeldrift.wordpress.com/2006/09/30/color-me-crazy-â-the-best-and-worst-paint-color-names/)
[4] http://search.dmoz.org/cgi-bin/search?search=Dodge+Charger

Dodge Charger

The **Dodge Charger** is an American → automobile manufactured by → Chrysler, under the → Dodge brand name. There have been several different Dodge vehicles, on three different platforms, bearing the Charger nameplate. The name is generally associated with a performance model in the Dodge range; however, it has also adorned a hatchback, a sedan, and a personal luxury coupe. The name was also carried by a 1999 concept car that differed substantially from the Charger eventually placed into production for the 2006 model year. A similar name, the Ramcharger, was used for the truck-based vehicle.

1969 Dodge Charger

Charger models

- 1964 Dodge Charger (1964 concept): a roadster-style show car based on the Dodge Polara
- 1965 Dodge Charger 273: a limited production option package for the Dart GT
- 1966–1978 → Dodge Charger (B-body): a rear wheel drive coupe and → muscle car
- 1983–1987 Dodge Charger (L-body): a front wheel drive subcompact hatchback
- 1999 Dodge Charger (1999 concept): a rear wheel drive concept car
- 2006–present Dodge Charger (LX): a rear wheel drive sedan that shares the same platform as the Chrysler 300, Dodge Magnum (discontinued), and current Dodge Challenger.

Other

- Dodge Charger Daytona - the name given to three different modified Dodge Charger's built on the → B-body and LX platforms.

See also

- 1971-1976 Chrysler Valiant Charger - short wheelbase Valiant coupe produced by Chrysler Australia
- The General Lee - Dodge Charger used in the television series *The Dukes of Hazzard*

External links

- Dodge Charger [1] - Information and history about the classic Dodge Charger.
- A history of the legendary Dodge Charger muscle car [2]

References

[1] http://www.stockmopar.com/dodge-charger.html
[2] http://www.allpar.com/model/charger.html

Car model

For scale models of automobiles, see Model car.

An **automobile model** (or **car model** or **model of car**, and typically abbreviated to just "model") is a particular brand of vehicle sold under a marque by a manufacturer, usually within a range of models, usually of different sizes or capabilities. From an engineering point of view, a particular car model is usually defined and/or constrained by the use of a particular car chassis/bodywork combination or the same monocoque, although sometimes this is not the case, and the model represents a marketing segment.

This engineering frame may have derivatives, giving rise to more than one body style for a particular car model. For example, the same model can be offered as a four-door sedan (saloon), a two-door coupé, a station wagon (estate), or even as a folding-roof convertible, all derived from essentially the same engineering frame. An example of this is the BMW 3-series.

Mechanical internals

The same car model can be offered with different mechanical internals, such as a choice of several engine sizes, automatic or manual transmissions, different suspension, braking or steering systems, etc.; all of these options considered fairly interchangeable on that specific body frame. It is common for any specific car model to carry additional badges or letterings to announce the mechanical option(s) incorporated on it.

However, when the same engineering body frame is sold under a different marque or by a partner automaker, it usually becomes, from a commercial point of view, a different car model. See badge engineering.

Marketing

Sometimes the marketing department may give each body style variant its own trade name, creating as many car models as body variants, even though they may share a large parts commonality and the engineering department may continue to consider them all part of the same project. An example of this is the Volkswagen Golf hatchback and the Volkswagen Jetta, which is of "three-box" design with a boot/trunk added to what is essentially a Golf. Conversely, the marketing department may advertise a car model as a convenient derivative of some popular car, when in fact they may be completely different engineering projects with almost no parts commonality, or from differing generations of the model. (For example, convertibles are often so heavily engineered, for a relatively small number of sales, that an older generation model is facelifted and carried forward with a new generation of the model's other body styles.)

Regional variations

The same car model may be sold by the automaker in different countries under different names. An example of this is the Mitsubishi Pajero / Montero.

Trim levels

A model may be offered in varying "trim levels", which usually affect little more than upholstery (cloth or leather, for example) and standard equipment. It is common for any specific car model to carry additional badges or letterings to announce its trim level. For example, the Toyota Camry's trim levels: Camry CE (Classic Edition), Camry LE (Luxury Edition), and Camry XLE (Extra Luxury Edition). Some manufacturers prefer names rather than initials for trim levels; the Renault Scénic range includes entry-level trim badged *Renault Scénic Authentique*, the next model up badged *Renault Scénic Expression*, then *Renault Scénic Dynamique* and finally the luxury *Renault Scénic Privilège*. The Mitsubishi Lancer Evolution's trim levels from Evolution I to Evolution X include: RS for Rally Sport; GSR for Grand Sport Rally; SE for Special Edition; MR for Mitsubishi Racing; GT-A for Grand Touring- Automatic; and FQ for Fucking Quick (note: FQ models sold in UK only).

The highest trim level is sometimes seen as slightly removed from the rest of the range. Ford traditionally have a *Ghia* luxury model above those which simply use initials, whilst Rover used the name of a former coachbuilder, *Vanden Plas*. There may also be a high-performance version such as a GT.

Market niches

Offering an array of body styles, mechanical specifications and trim levels allow manufacturers to target the same car model to different market niches. For example, the cheap, basic-trim-level, three-door variant of some popular car may be right for the student on a budget, while the station wagon with comfort package may suit the needs of an elder lady, and the very expensive, high-performance, semi-racing variant may catch the eye of the sportier-minded executive with a fat wallet, all of the three variants having arisen from the same project and carrying the same commercial name. An example of this is the Ford Focus.

In a trim, terms like LX, EX, etc are also referred to as Grade.

Model years

A car model may be further subdivided into model years, all cars from a particular model year sharing approximately the same characteristics (given the same trim level, body style, engine option, etc.) but sometimes with slight differences from others of a different model year. In this context, a face lift may be used to slightly update the looks of an aging car model without a major engineering revision, giving way to a so-called "second series" of that particular model, and sometimes becoming the opportunity for a marketing re-launch of the same car.

Many times a manufacturer decides to completely redesign the car, but with the aim of offering the new model to the same specific public or in the same market niche, keeping it similarly priced and marketed against its usual competitors from other manufacturers. The car is usually considered a different model by the engineering department, carrying a different model designator, but, for marketing reasons, it is offered to the consumers with the same old, traditional, familiar name. An example of this is the Chevrolet Corvette.

Total production run for a given car is usually calculated regarding the engineering project name or designator. The marketing department may advertise figures for a continuous-production tradename instead, divided in so-called "generations". However, for government or sport regulatory purposes, each body-style/mechanical-configuration combination may be counted as a different model.

See also

- Automaker
- Facelift (automobile)
- Marque
- Automotive package
- Restyling
- Model year
- List of automobile model and marque oddities
- Badge engineering

Automobile

An **automobile**, **motor car** or **car** is a wheeled motor vehicle used for transporting passengers, which also carries its own engine or motor. Most definitions of the term specify that automobiles are designed to run primarily on roads, to have seating for one to eight people, to typically have four wheels, and to be constructed principally for the transport of people rather than goods.[1] However, the term *automobile* is far from precise, because there are many types of vehicles that do similar tasks.

As of 2002, there were 590 million passenger cars worldwide (roughly one car per eleven people).[2] Around the world, there were about 806 million cars and light trucks on the road in 2007; they burn over 260 billion gallons of gasoline and diesel fuel yearly. The numbers are increasing rapidly, especially in China and India.[3]

Karl Benz's "Velo" model (1894) - entered into an early automobile race

Passenger cars in 2000

Etymology

The word **automobile** comes, via the French *automobile*, from the Ancient Greek word αὐτός (*autós*, "self") and the Latin *mobilis* ("movable"); meaning a vehicle that moves itself, rather than being pulled or pushed by a separate animal or another vehicle. The alternative name *car* is believed to originate from the Latin word *carrus* or *carrum* ("wheeled vehicle"), or the Middle English word *carre* ("cart") (from Old North French), or *karros* (a Gallic wagon).[4] [5]

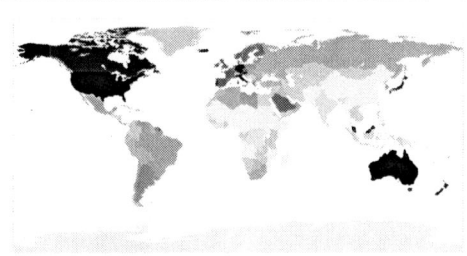

World map of passenger cars per 1000 people.

History

Ferdinand Verbiest, a member of a Jesuit mission in China, built the first steam-powered vehicle around 1672 which was of small scale and designed as a toy for the Chinese Emperor, that was unable to carry a driver or a passenger, but quite possibly, was the first working steam-powered vehicle ('auto-mobile').[6] [7]

Although Nicolas-Joseph Cugnot is often credited with building the first self-propelled mechanical vehicle or automobile in about 1769, by adapting an existing horse-drawn vehicle, this claim is disputed by some, who doubt Cugnot's three-wheeler ever ran or was stable. What is not in doubt is that Richard Trevithick built and demonstrated his *Puffing Devil* road locomotive in 1801, believed by many to be the first demonstration of a steam-powered road vehicle, although it was unable to maintain sufficient steam pressure for long periods, and would have been of little practical use.

In Russia, in the 1780s, Ivan Kulibin developed a human-pedalled, three-wheeled carriage with modern features such as a flywheel, brake, gear box, and bearings; however, it was not developed further.[8]

François Isaac de Rivaz, a Swiss inventor, designed the first internal combustion engine, in 1806, which was fueled by a mixture of hydrogen and oxygen and used it to develop the world's first vehicle, albeit rudimentary, to be powered by such an engine. The design was not very successful, as was the case with others, such as Samuel Brown, Samuel Morey, and Etienne Lenoir with his hippomobile, who each produced vehicles (usually adapted carriages or carts) powered by clumsy internal combustion engines.[9]

In November 1881, French inventor Gustave Trouvé demonstrated a working three-wheeled automobile that was powered by electricity. This was at the International Exhibition of Electricity in Paris.[10]

Although several other German engineers (including Gottlieb Daimler, Wilhelm Maybach, and Siegfried Marcus) were working on the problem at about the same time, **Karl Benz** generally is acknowledged as the inventor of the modern automobile.[9]

An automobile powered by his own four-stroke cycle gasoline engine was built in Mannheim, Germany by Karl Benz in 1885, and granted a patent in January of the following year under the auspices of his major company, Benz & Cie., which was founded in 1883. It was an integral design, without the adaptation of other existing components, and included several new technological elements to create a new concept. This is what made it worthy of a patent. He began to sell his production vehicles in 1888.

Karl Benz

A photograph of the original *Benz Patent Motorwagen*, first built in 1885 and awarded the patent for the concept

In 1879, Benz was granted a patent for his first engine, which had been designed in 1878. Many of his other inventions made the use of the internal combustion engine feasible for powering a vehicle.

His first *Motorwagen* was built in 1885, and he was awarded the patent for its invention as of his application on January 29, 1886. Benz began promotion of the vehicle on July 3, 1886, and about 25 Benz vehicles were sold between 1888 and 1893, when his first four-wheeler was introduced along with a model intended for affordability. They also were powered with four-stroke engines of his own design. Emile Roger of France, already producing Benz engines under license, now added the Benz automobile to his line of products. Because France was more open to the early automobiles, initially more were built and sold in France through Roger than Benz sold in Germany.

In 1896, Benz designed and patented the first internal-combustion flat engine, called a *boxermotor* in German. During the last years of the nineteenth century, Benz was the largest automobile company in the world with 572 units produced in 1899 and, because of its size, Benz & Cie., became a joint-stock company.

Daimler and Maybach founded Daimler Motoren Gesellschaft (Daimler Motor Company, DMG) in Cannstatt in 1890, and under the brand name, *Daimler*, sold their first automobile in 1892, which was a horse-drawn stagecoach built by another manufacturer, that they retrofitted with an engine of their design. By 1895 about 30 vehicles had been built by Daimler and Maybach, either at the Daimler works or in the Hotel Hermann, where they set up shop after disputes with their backers. Benz and the Maybach and the Daimler team seem to have been unaware of each other's early work. They never worked together because, by the time of the merger of the two companies, Daimler and Maybach were no longer part of DMG.

Daimler died in 1900 and later that year, Maybach designed an engine named *Daimler-Mercedes*, that was placed in a specially-ordered model built to specifications set by Emil Jellinek. This was a production of a small number of vehicles for Jellinek to race and market in his country. Two years later, in 1902, a new model DMG automobile was produced and the model was named Mercedes after the Maybach engine which generated 35 hp. Maybach quit DMG shortly thereafter and opened a business of his own. Rights to the *Daimler* brand name were sold to other manufacturers.

Karl Benz proposed co-operation between DMG and Benz & Cie. when economic conditions began to deteriorate in Germany following the First World War, but the directors of DMG refused to consider it initially. Negotiations between the two companies resumed several years later when these conditions worsened and, in 1924 they signed an *Agreement of Mutual Interest*, valid until the year 2000. Both enterprises standardized design, production, purchasing, and sales and they advertised or marketed their automobile models jointly, although keeping their respective brands.

On June 28, 1926, Benz & Cie. and DMG finally merged as the *Daimler-Benz* company, baptizing all of its automobiles *Mercedes Benz*, as a brand honoring the most important model of the DMG automobiles, the Maybach design later referred to as the *1902 Mercedes-35 hp*, along with the Benz name. Karl Benz remained a member of the board of directors of Daimler-Benz until his death in 1929, and at times, his two sons participated in the management of the company as well.

In 1890, Émile Levassor and Armand Peugeot of France began producing vehicles with Daimler engines, and so laid the foundation of the automobile industry in France.

The first design for an American automobile with a gasoline internal combustion engine was drawn in 1877 by George Selden of Rochester, New York, who applied for a patent for an automobile in 1879, but the patent application expired because the vehicle was never built. After a delay of sixteen years and a series of attachments to his application, on November 5, 1895, Selden was granted a United States patent (U.S. Patent 549160 [11]) for a two-stroke automobile engine, which hindered, more than encouraged, development of automobiles in the United States. His patent was challenged by Henry Ford and others, and overturned in 1911.

In Britain, there had been several attempts to build steam cars with varying degrees of success, with Thomas Rickett even attempting a production run in 1860.[12] Santler from Malvern is recognized by the Veteran Car Club of Great Britain as having made the first petrol-powered car in the country in 1894[13] followed by Frederick William Lanchester in 1895, but these were both one-offs.[13] The first production vehicles in Great Britain came from the Daimler Motor Company, a company founded by Harry J. Lawson in 1896, after purchasing the right to use the name of the engines. Lawson's company made its first automobiles in 1897, and they bore the name *Daimler*.[13]

In 1892, German engineer Rudolf Diesel was granted a patent for a "New Rational Combustion Engine". In 1897, he built the first Diesel Engine.[9] Steam-, electric-, and gasoline-powered vehicles competed for decades, with gasoline internal combustion engines achieving dominance in the 1910s.

Although various pistonless rotary engine designs have attempted to compete with the conventional piston and crankshaft design, only Mazda's version of the Wankel engine has had more than very limited success.

Production

The large-scale, production-line manufacturing of affordable automobiles was debuted by Ransom Olds at his Oldsmobile factory in 1902. This concept was greatly expanded by Henry Ford, beginning in 1914.

As a result, Ford's cars came off the line in fifteen minute intervals, much faster than previous methods, increasing productivity eightfold (requiring 12.5 man-hours before, 1 hour 33 minutes after), while using less manpower.[14] It was so successful, paint became a bottleneck. Only Japan black would dry fast enough, forcing the company to drop the variety of colors available before 1914, until fast-drying Duco lacquer was developed in 1926. This is the source of Ford's apocryphal remark, "any color as long as it's black".[14] In 1914, an assembly line worker could buy a Model T with four months' pay.[14]

Ransom E. Olds.

Portrait of Henry Ford (ca. 1919)

Ford's complex safety procedures—especially assigning each worker to a specific location instead of allowing them to roam about—dramatically reduced the rate of injury. The combination of high wages and high efficiency is called "Fordism," and was copied by most major industries. The efficiency gains from the assembly line also coincided with the economic rise of the United States. The assembly line forced workers to work at a certain pace with very repetitive motions which led to more output per worker while other countries were using less productive methods.

In the automotive industry, its success was dominating, and quickly spread worldwide seeing the founding of Ford France and Ford Britain in 1911, Ford Denmark 1923, Ford Germany 1925; in 1921, Citroen was the first native European manufacturer to adopt the production method. Soon, companies had to have assembly lines, or risk going broke; by 1930, 250 companies which did not, had disappeared.[14]

Development of automotive technology was rapid, due in part to the hundreds of small manufacturers competing to gain the world's attention. Key developments included electric ignition and the electric self-starter (both by Charles Kettering, for the Cadillac Motor Company in 1910-1911), independent suspension, and four-wheel brakes.

Since the 1920s, nearly all cars have been mass-produced to meet market needs, so marketing plans often have heavily influenced automobile design. It was Alfred P. Sloan who established the idea of different makes of cars produced by one company, so buyers could "move up" as their fortunes improved.

Ford Model T, 1927, regarded as the first affordable American automobile

Reflecting the rapid pace of change, makes shared parts with one another so larger production volume resulted in lower costs for each price range. For example, in the 1930s, LaSalles, sold by Cadillac, used cheaper mechanical parts made by Oldsmobile; in the 1950s, Chevrolet shared hood, doors, roof, and windows with Pontiac; by the 1990s, corporate drivetrains and shared platforms (with interchangeable brakes, suspension, and other parts) were common. Even so, only major makers could afford high costs, and even companies with decades of production, such as Apperson, Cole, Dorris, Haynes, or Premier, could not manage: of some two hundred American car makers in existence in 1920, only 43 survived in 1930, and with the Great Depression, by 1940, only 17 of those were left.[14]

In Europe much the same would happen. Morris set up its production line at Cowley in 1924, and soon outsold Ford, while beginning in 1923 to follow Ford's practise of vertical integration, buying Hotchkiss (engines), Wrigley (gearboxes), and Osberton (radiators), for instance, as well as competitors, such as Wolseley: in 1925, Morris had 41% of total British car production. Most British small-car assemblers, from Abbey to Xtra had gone under. Citroen did the same in France, coming to cars in 1919; between them and other cheap cars in reply such as Renault's 10CV and Peugeot's 5CV, they produced 550,000 cars in 1925, and Mors, Hurtu, and others could not compete.[14] Germany's first mass-manufactured car, the Opel 4PS *Laubfrosch* (Tree Frog), came off the line at Russelsheim in 1924, soon making Opel the top car builder in Germany, with 37.5% of the market.[14]

Fuel and propulsion technologies

Most automobiles in use today are propelled by gasoline (also known as petrol) or diesel internal combustion engines, which are known to cause air pollution and are also blamed for contributing to climate change and global warming.[15] Increasing costs of oil-based fuels, tightening environmental laws and restrictions on greenhouse gas emissions are propelling work on alternative power systems for automobiles. Efforts to improve or replace existing technologies include the development of hybrid vehicles, and electric and hydrogen vehicles which do not release pollution into the air.

Petroleum fuels

Diesel

Diesel-engined cars have long been popular in Europe with the first models being introduced as early as 1922 [16] by Peugeot and the first production car, Mercedes-Benz 260 D in 1936 by Mercedes-Benz. The main benefit of diesel engines is a 50% fuel burn efficiency compared with 27%[17] in the best gasoline engines. A down-side of the Diesel engine is that better filters are required to reduce the presence in the exhaust gases of fine soot particulates called diesel particulate matter. Manufacturers are now starting to fit diesel particulate filters to remove the soot. Many diesel-powered cars can run with little or no modifications on 100% biodiesel and combinations of other organic oils.

A radio taxi in New Delhi. A court order requires all commercial vehicles including trucks, buses and taxis in India to run on Compressed Natural Gas

Gasoline

Gasoline engines have the advantage over diesel in being lighter and able to work at higher rotational speeds and they are the usual choice for fitting in high-performance sports cars. Continuous development of gasoline engines for over a hundred years has produced improvements in efficiency and reduced pollution. The carburetor was used on nearly all road car engines until the 1980s but it was long realised better control of the fuel/air mixture could be achieved with fuel injection. Indirect fuel injection was first used in aircraft engines from 1909, in racing car engines from the 1930s, and road cars from the late 1950s.[17] Gasoline Direct Injection (GDI) is now starting to appear in production vehicles such as the 2007 (Mark II) BMW Mini. Exhaust gases are also cleaned up by fitting a catalytic converter into the exhaust system. Clean air legislation in many of the car industries most important markets has made both catalysts and fuel injection virtually universal fittings. Most modern gasoline engines also are capable of running with up to 15% ethanol mixed into the gasoline - older vehicles may have seals and hoses that can be harmed by ethanol. With a small amount of redesign, gasoline-powered vehicles can run on ethanol concentrations as high as 85%. 100% ethanol is used in some parts of the world (such as Brazil), but vehicles must be started on pure gasoline and switched over to ethanol once the engine is running. Most gasoline engined cars can also run on LPG with the addition of an LPG tank for fuel storage and carburettor modifications to add an LPG mixer. LPG produces fewer toxic emissions and is a popular fuel for fork-lift trucks that have to operate inside buildings.

2007 Mark II (BMW) Mini Cooper

Biofuels

Ethanol, other alcohol fuels (biobutanol) and biogasoline have widespread use as an automotive fuel. Most alcohols have less energy per liter than gasoline and are usually blended with gasoline. Alcohols are used for a variety of reasons - to increase octane, to improve emissions, and as an alternative to petroleum based fuel, since they can be made from agricultural crops. Brazil's ethanol program provides about 20% of the nation's automotive fuel needs, as a result of the mandatory use of E25 blend of gasoline throughout the country, 3 million cars that operate on pure ethanol, and 6 million dual or flexible-fuel vehicles sold since 2003,[18] that run on any mix of ethanol and gasoline. The commercial success of "flex" vehicles, as they are popularly known, have allowed sugarcane based ethanol fuel to achieve a 50% market share of the gasoline market by April 2008.[19] [20] [21]

The hydrogen powered FCHV (Fuel Cell Hybrid Vehicle) was developed by Toyota in 2005

Electric

The first electric cars were built around 1832, well before internal combustion powered cars appeared.[22] For a period of time electrics were considered superior due to the silent nature of electric motors compared to the very loud noise of the gasoline engine. This advantage was removed with Hiram Percy Maxim's invention of the muffler in 1897. Thereafter internal combustion powered cars had two critical advantages: 1) long range and 2) high specific energy (far lower weight of petrol fuel versus weight of batteries). The building of battery electric vehicles that could rival internal combustion models had to wait for the introduction of modern semiconductor controls and improved batteries. Because they can deliver a high torque at low revolutions electric cars do not require such a complex drive train and transmission as internal combustion powered cars. Some post-2000 electric car designs such as the Venturi Fétish are able to accelerate from 0-60 mph (96 km/h) in 4.0 seconds with a top speed around 130 mph (210 km/h). Others have a range of 250 miles (400 km) on the United States Environmental Protection Agency (EPA) highway cycle requiring 3-1/2 hours to completely charge.[23] Equivalent fuel efficiency to internal combustion is not well defined but some press reports give it at around 135 miles per US gallon (1.74 L/100 km; 162 mpg_{-imp}).

The Henney Kilowatt, the first modern (transistor-controlled) electric car.

2007 Tesla electric powered Roadster

Hydrogen

Hydrogen is a fuel that, upon consumption, does not emit any greenhouse gases. Hydrogen can be burned in internal combustion engines as well as fuel cells.

Oxyhydrogen

Oxyhydrogen is another fuel that can be used in existing internal combustion engines originally developed for using gasoline. This allows the engine to eliminate emissions, although fuel efficiency is reduced rather than improved (since the energy required to split water exceeds the energy recouped by burning it).

Tata/MDI OneCAT Air Car

Steam

Steam power, usually using an oil- or gas-heated boiler, was also in use until the 1930s but had the major disadvantage of being unable to power the car until boiler pressure was available (although the newer models could achieve this in well under a minute). It has the advantage of being able to produce very low emissions as the combustion process can be carefully controlled. Its disadvantages include poor heat efficiency and extensive requirements for electric auxiliaries.[24].

Air

A CNG powered high-floor Neoplan AN440A, run on Compressed Natural Gas

A compressed air car is an alternative fuel car that uses a motor powered by compressed air. The car can be powered solely by air, or by air combined (as in a hybrid electric vehicle) with gasoline/diesel/ethanol or electric plant and regenerative braking. Instead of mixing fuel with air and burning it to drive pistons with hot expanding gases; *compressed air cars* use the expansion of compressed air to drive their pistons. Several prototypes are available already and scheduled for worldwide sale by the end of 2008, though this has not happened as of January 2009. Companies releasing this type of car include Tata Motors and Motor Development International (MDI).

Gas turbine

In the 1950s there was a brief interest in using gas turbine engines and several makers including Rover and → Chrysler produced prototypes. In spite of the power units being very compact, high fuel consumption, severe delay in throttle response, and lack of engine braking meant no cars reached production.

Rotary (Wankel) engines

Rotary Wankel engines were introduced into road cars by NSU with the Ro 80 and later were seen in the Citroën GS Birotor and several Mazda models. In spite of their impressive smoothness, poor reliability and fuel economy has largely lead to their decline. Mazda, beginning with the R100 then RX-2, has continued research on these engines, overcoming most of the earlier problems with the RX-7 and RX-8.

Rocket and jet cars

A rocket car holds the record in drag racing. However, the fastest of those cars are used to set the Land Speed Record, and are propelled by propulsive jets emitted from rocket, turbojet, or more recently and most successfully turbofan engines. The ThrustSSC car using two Rolls-Royce Spey turbofans with reheat was able to exceed the speed of sound at ground level in 1997.

Data transmission

Automobiles use CAM, MOSH (optic fiber), multiplexing, bluetooth and WiFi between others.

Safety

There are three main statistics to which automobile safety can be compared:[25]

Result of a serious automobile accident.

Deaths per billion journeys
Bus: 4.3
Rail: 20
Van: 20
Car: 40
Foot: 40
Water: 90
Air: 117
Bicycle: 170
Motorcycle: 1640

Deaths per billion hours
Bus: 11.1
Rail: 30
Air: 30.8
Water: 50
Van: 60
Car: 130
Foot: 220
Bicycle: 550
Motorcycle: 4840

Deaths per billion kilometres
Air: 0.05
Bus: 0.4
Rail: 0.6
Van: 1.2
Water: 2.6
Car: 3.1
Bicycle: 44.6
Foot: 54.2
Motorcycle: 108.9

While road traffic injuries represent the leading cause in worldwide injury-related deaths,[26] their popularity undermines this statistic.

Mary Ward became one of the first documented automobile fatalities in 1869 in Parsonstown, Ireland[27] and Henry Bliss one of the United States' first pedestrian automobile casualties in 1899 in New York.[28] There are now standard tests for safety in new automobiles, like the EuroNCAP and the US NCAP tests,[29] as well as insurance-backed IIHS tests.[30]

Costs and benefits

The costs of automobile usage, which may include the cost of: acquiring the vehicle, repairs, maintenance, fuel, depreciation, parking fees, tire replacement, taxes and insurance,[31] are weighed against the cost of the alternatives, and the value of the benefits - perceived and real - of vehicle usage. The benefits may include on-demand transportation, mobility, independence and convenience.[7]

Similarly the costs to society of encompassing automobile use, which may include those of: maintaining roads, land use, pollution, public health, health care, and of disposing of the vehicle at the end of its life, can be balanced against the value of the benefits to society that automobile use generates. The societal benefits may include: economy benefits, such as job and wealth creation, of automobile production and maintenance, transportation provision, society wellbeing derived from leisure and travel opportunities, and revenue generation from the tax opportunities. The ability for humans to move flexibly from place to place has far reaching implications for the nature of societies.[32]

Environmental impact

Transportation is a major contributor to air pollution in most industrialised nations. According to the American Surface Transportation Policy Project nearly half of all Americans are breathing unhealthy air. Their study showed air quality in dozens of metropolitan areas has worsened over the last decade.[33] In the United States the average passenger car emits 11,450 lbs (5 tonnes) of carbon dioxide, along with smaller amounts of carbon monoxide, hydrocarbons, and nitrogen.[34]

Animals and plants are often negatively impacted by automobiles via habitat destruction and pollution. Over the lifetime of the average automobile the "loss of habitat potential" may be over 50,000 square meters (538,195 square feet) based on Primary production correlations.[35]

Fuel taxes may act as an incentive for the production of more efficient, hence less polluting, car designs (e.g. hybrid vehicles) and the development of alternative fuels. High fuel taxes may provide a strong incentive for consumers to purchase lighter, smaller, more fuel-efficient cars, or to not drive. On average, today's automobiles are about 75 percent recyclable, and using recycled steel helps reduce energy use and pollution.[36] In the United States Congress, federally mandated fuel efficiency standards have been debated regularly, passenger car standards have not risen above the 27.5 miles per US gallon (8.55 L/100 km; 33.0 mpg_{-imp}) standard set in 1985. Light truck standards have changed more frequently, and were set at 22.2 miles per US gallon (10.6 L/100 km; 26.7 mpg_{-imp}) in 2007.[37] Alternative fuel vehicles are another option that is less polluting than conventional petroleum powered vehicles.

Other negative effects

Residents of low-density, residential-only sprawling communities are also more likely to die in car collisions which kill 1.2 million people worldwide each year, and injure about forty times this number.[26] Sprawl is more broadly a factor in inactivity and obesity, which in turn can lead to increased risk of a variety of diseases.[38]

Driverless cars

Fully autonomous vehicles, also known as robotic cars, or driverless cars, already exist in prototype, and are expected to be commercially available around 2020. According to urban designer and futurist Michael E. Arth, driverless electric vehicles—in conjunction with the increased use of virtual reality for work, travel, and pleasure—could reduce the world's 800,000,000 vehicles to a fraction of that number within a few decades.[39] This would be possible if almost all private cars requiring drivers, which are not in use and parked 90% of the time, would be traded for public self-driving taxis that would be in near constant use. This would also allow for getting the appropriate vehicle for the particular need—a bus could come for a group of people, a limousine could come for a special night out, and a Segway could come for a short trip down the street for one person. Children could be chauffeured in supervised safety, DUIs would no longer exist, and 41,000 lives could be saved each year in the U.S. alone.[40] [41]

Future car technologies

Automobile propulsion technology under development include gasoline/electric and plug-in hybrids, battery electric vehicles, hydrogen cars, biofuels, and various alternative fuels.

Research into future alternative forms of power include the development of fuel cells, Homogeneous Charge Compression Ignition (HCCI), stirling engines[42], and even using the stored energy of compressed air or liquid nitrogen.

New materials which may replace steel car bodies include duraluminum, fiberglass, carbon fiber, and carbon nanotubes.

Telematics technology is allowing more and more people to share cars, on a pay-as-you-go basis, through such schemes as City Car Club in the UK, Mobility in mainland Europe, and Zipcar in the US.

Alternatives to the automobile

Established alternatives for some aspects of automobile use include public transit (buses, trolleybuses, trains, subways, monorails, tramways), cycling, walking, rollerblading, skateboarding, horseback riding and using a velomobile. Car-share arrangements and carpooling are also increasingly popular–the U.S. market leader in car-sharing has experienced double-digit growth in revenue and membership growth between 2006 and 2007, offering a service that enables urban residents to "share" a vehicle rather than own a car in already congested neighborhoods.[43] Bike-share systems have been tried in some European cities, including Copenhagen and Amsterdam. Similar programs have been experimented with in a number of U.S. Cities.[44] Additional individual modes of transport, such as personal rapid transit could serve as an alternative to automobiles if they prove to be socially accepted.[45]

See also

- Air pollution
- Bus
- Car classification
- Carfree city
- Driving
- List of countries by automobile production
- List of countries by vehicles per capita
- Lists of automobiles
- Motor vehicle theft
- Noise pollution
- Steering
- Society of Automotive Engineers
- Sustainable transport
- Traffic collision
- Traffic congestion
- Truck
- U.S. Automobile Production Figures - production figures for each make from 1899 to 2000

Further reading

- Halberstam, David, *The Reckoning*, New York, Morrow, 1986. ISBN 0688048382
- Kay, Jane Holtz, *Asphalt nation : how the automobile took over America, and how we can take it back*, New York, Crown, 1997. ISBN 0517587025
- Heathcote Williams, *Autogeddon*, New York, Arcade, 1991. ISBN 1559701765

External links

- Fédération Internationale de l'Automobile [46]
- Forum for the Automobile and Society [47]

References

[1] compiled by F.G. Fowler and H.W. Fowler. (1976). *Pocket Oxford Dictionary*. London: Oxford University Press. ISBN 0-19-861113-7.
[2] " WorldMapper - passenger cars (http://www.sasi.group.shef.ac.uk/worldmapper/display.php?selected=31)". .
[3] Plunkett Research, "Automobile Industry Introduction" (2008) (http://www.plunkettresearch.com/Industries/AutomobilesTrucks/AutomobileTrends/tabid/89/Default.aspx)
[4] " "Car" (http://www.etymonline.com/index.php?term=car)". *(etymology)*. Online Etymology Dictionary. . Retrieved 2008-06-02.
[5] (http://www.lib.wayne.edu/resources/special_collections/local/cfai/index.php), 'Car' derived from 'carrus'.
[6] " 1679-1681–R P Verbiest's Steam Chariot (http://translate.google.com/translate?hl=en&sl=fr&u=http://users.skynet.be/tintinpassion/VOIRSAVOIR/Auto/Pages_auto/Auto_001.html&sa=X&oi=translate)". *History of the Automobile: origin to 1900*. Hergé. . Retrieved 2009-05-08.
[7] Setright, L. J. K. (2004). *Drive On!: A Social History of the Motor Car*. Granta Books. ISBN 1-86207-698-7.
[8] " Automobile Invention (http://www.aboutmycar.com/category/car_history/creation_history/automobile-invention-1122.htm)". Aboutmycar.com. . Retrieved 2008-10-27.
[9] Ralph Stein (1967). *The Automobile Book*. Paul Hamlyn Ltd.
[10] Wakefield, Ernest H. (1994). *History of the Electric Automobile*. Society of Automotive Engineers, Inc.. p. 2–3. ISBN 1-56091-299-5.
[11] http://www.google.com/patents?vid=549160
[12] Burgess Wise, D. (1970). *Veteran and Vintage Cars*. London: Hamlyn. ISBN 0-600-00283-7.
[13] Georgano, N. (2000). *Beaulieu Encyclopedia of the Automobile*. London: HMSO. ISBN 1-57958-293-1.
[14] Georgano, G. N. (2000). *Vintage Cars 1886 to 1930*. Sweden: AB Nordbok. ISBN 1-85501-926-4.
[15] " Global Climate Change (http://www.fueleconomy.gov/feg/climate.shtml)". U.S. Department of Energy. . Retrieved 2007-03-03.
[16] " Economical Lion Brand Cars (http://www.peugeot.com/en/history/a-century-of-expertise/economical-lion-brand-cars.aspx)". Peugeot Automobiles. 2009-03-13. . Retrieved 2009-03-13.
[17] Norbye, Jan (1988). *Automotive fuel injection Systems*. Haynes Publishing. ISBN 0-85429-755-3.
[18] " Veículos flex somam 6 milhões e alcançam 23% da frota (http://www1.folha.uol.com.br/folha/dinheiro/ult91u428265.shtml)" (in Portuguese). Folha Online. 2008-08-04. . Retrieved 2008-08-04.
[19] Agência Brasil (2008-08-15). " ANP: consumo de álcool combustível é 50% maior em 2007 (http://br.invertia.com/noticias/noticia.aspx?idNoticia=200807152306_ABR_77211977)" (in Portuguese). Invertia. . Retrieved 2008-08-09.
[20] Gazeta Mercantil (2008). " ANP estima que consumo de álcool supere gasolina (http://www.agropecuariabrasil.com.br/anp-estima-que-consumo-de-alcool-supere-gasolina/)" (in Portuguese). Agropecuária Brasil. . Retrieved 2008-08-09.
[21] Inslee, Jay; Bracken Hendricks (2007), *Apollo's Fire*, Island Press, Washington, D.C., pp. 153–155, 160–161, ISBN 978-1-59726-175-3 . See Chapter 6. Homegrown Energy.
[22] Bellis, M. (2006) "The History of Electric Vehicles: The Early Years" *About.com* article at inventors.about.com (http://inventors.about.com/library/weekly/aacarselectrica.htm) accessed on 5 September 2007
[23] Mitchell, T. (2003) "AC Propulsion Debuts tzero with LiIon Battery" *AC Propulsion, Inc.* press release at acpropulsion.com (http://web.archive.org/web/20080307073527/http://www.acpropulsion.com/LiIon_tzero_release.pdf) accessed 5 September 2007
[24] Setright, L.J.K. "Steam: The Romantic Illusion", in Ward, Ian, ed., *World of Automobiles* (London: Orbis Publishing, 1974), pp.2168-2173.)
[25] " The risks of travel (http://www.numberwatch.co.uk/risks_of_travel.htm)". Numberwatch.co.uk. . Retrieved 2008-10-27.
[26] Peden M, Scurfield R, Sleet D *et al.* (eds.) (2004). *World report on road traffic injury prevention* (http://who.int/violence_injury_prevention/publications/road_traffic/world_report/en/). World Health Organization. ISBN 92-4-156260-9. . Retrieved 2008-06-24.
[27] " Mary Ward 1827–1869 (http://www.universityscience.ie/pages/scientists/sci_mary_ward.php)". Universityscience.ie. . Retrieved 2008-10-27.
[28] " CityStreets - Bliss plaque (http://www.citystreets.org/plaque.html)". .
[29] " SaferCar.gov - NHTSA (http://www.nhtsa.dot.gov/cars/testing/ncap/)". .
[30] " Insurance Institute for Highway Safety (http://www.hwysafety.org/)". .
[31] " car operating costs (http://www.racv.com.au/wps/wcm/connect/Internet/Primary/my+car/advice+&+information/car+operating+costs/)". *my car*. RACV. . Retrieved 2006-12-01.
[32] John A. Jakle, Keith A. Sculle. (2004). *Lots of Parking: Land Use in a Car Culture*. Charlottesville: Univ. of Virginia Press. ISBN 0813922666.
[33] " Clearing the Air (http://www.transact.org/report.asp?id=227)". The Surface Transportation Policy Project. 2003-08-19. . Retrieved 2007-04-26.
[34] " Emission Facts (http://www.epa.gov/otaq/consumer/f00013.htm)". United States Environmental Protection Agency. .
[35] " ecological effects of automobiles (http://ecofx.org/wiki/index.php?title=Automobiles)". ecofx. .
[36] " Automobiles and the Environment (http://web.archive.org/web/20080214145812/http://www.greenercars.com/autoenviron.html)". Greenercars.com. Archived from the original (http://www.greenercars.com/autoenviron.html) on 2008-02-14. .
[37] ;" CAFE Overview - Frequently Asked Questions (http://www.nhtsa.dot.gov/cars/rules/cafe/overview.htm)". National Highway Traffic Safety Administration. .
[38] " Our Ailing Communities (http://www.metropolismag.com/cda/story.php?artid=2353)". Metropolis Magazine. .

[39] Oliver, Rachel (2007-09-16). " Rachel Oliver "All About: hydrid transportation" (http://www.cnn.com/2007/BUSINESS/09/14/allabout.hybrid/)". CNN. . Retrieved 2009-03-05.
[40] Arth, Michael (Spring 2008). " "New Pedestrianism: A Bridge to the Future" (http://www.carbusters.org/magazine/33/feature3.html)". Carbusters Magazine. . Retrieved 2009-03-06.
[41] Birch, Alex (2008-05-23). " "Most Cars Can be Eliminated in 20 Years says Urban Designer Michael E. Arth" (http://www.corrupt.org/news/most_cars_can_be_eliminated_in_20_years_says_urban_designer_michael_e_arth)". Corrupt.org. . Retrieved 2009-03-06.
[42] Paul Werbos. " Who Killed the Electric car? My review (http://www.werbos.com/E/WhoKilledElecPJW.htm)". . Retrieved 2007-04-10.
[43] " Flexcar Expands to Philadelphia (http://www.greencarcongress.com/2007/04/flexcar_expands.html)". Green Car Congress. 2007-04-02. .
[44] " About Bike Share Programs (http://web.archive.org/web/20071220235050/http://web.mit.edu/dzshen/www/about.shtml)". Tech Bikes MIT. Archived from the original (http://web.mit.edu/dzshen/www/about.shtml) on 2007-12-20. .
[45] Jane Holtz Kay (1998). *Asphalt Nation: how the automobile took over America, and how we can take it back*. Berkeley, Calif.: University of California Press. ISBN 0520216202.
[46] http://www.fia.com/
[47] http://www.autoandsociety.com/

Dodge

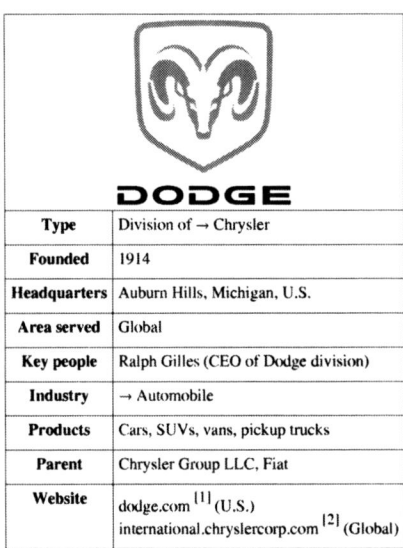

Type	Division of → Chrysler
Founded	1914
Headquarters	Auburn Hills, Michigan, U.S.
Area served	Global
Key people	Ralph Gilles (CEO of Dodge division)
Industry	→ Automobile
Products	Cars, SUVs, vans, pickup trucks
Parent	Chrysler Group LLC, Fiat
Website	dodge.com [1] (U.S.) international.chryslercorp.com [2] (Global)

Dodge is a United States-based brand of → automobiles, minivans, sport utility vehicles, and pickup trucks, manufactured and marketed by Chrysler Group LLC in more than 60 different countries and territories worldwide. Founded as the Dodge Brothers Company in 1900 to supply parts and assemblies for Detroit's growing auto industry, Dodge began making its own complete vehicles in 1914. The brand was sold to Chrysler Corporation in 1928, passed through the short-lived DaimlerChrysler merger of 1998–2007 as part of the Chrysler Group, was a part of **Chrysler LLC** owned by Cerberus Capital Management, a private equity investment firm, and is now a part of the **Chrysler Group LLC** which has an alliance with Fiat. Fiat has plans to eliminate some Dodge, Chrysler, and Jeep existing vehicles in favor of Fiat-Chrysler co-developed vehicles. [3] [4]

Founding and early years

After the founding of the Dodge Brothers Company by Horace and John Dodge in 1900, the Detroit-based company quickly found work producing precision engine and chassis components for the city's burgeoning number of automobile firms. Chief among these customers were the established Olds Motor Vehicle Company and the then-new Ford Motor Company. Dodge Brothers enjoyed much success in this field, but the brothers' growing wish to build complete vehicles was exemplified by John Dodge's 1913 exclamation that he was "tired of being carried around in Henry Ford's vest pocket."

1915 **Dodge Brothers** Model 30-35 touring car

By 1914, he and Horace had fixed that by creating the new four-cylinder Dodge Model 30. Pitched as a slightly more upscale competitor to the ubiquitous Ford Model T, it pioneered or made standard many features later taken for granted: all-steel body construction (when the vast majority of cars worldwide still used wood framing under steel panels, though Stoneleigh and BSA had used steel bodies as early as 1911),[5] 12-volt electrical system (6-volt systems would remain the norm up until the 1950s), and sliding-gear transmission (the best-selling Model T would retain an antiquated planetary design all the way until its demise in 1927). As a result of all this, as well as the brothers' well-earned reputation for quality through the parts they had made for other successful vehicles, Dodge Brothers cars were ranked at second place for U.S. sales as early as 1916. The same year, Henry Ford decided to stop paying dividends to finance the construction of the new River Rouge complex, leading to the Dodge brothers filing suit to protect approximately a million dollars a year they were earning;[6] this led Ford to buy out his shareholders, and the Dodges were paid some US$25 million.[7]

Dodge Brothers delivery trucks, Salt Lake City, 1920

In the same year, Dodge Brothers vehicles won wide acclaim for durability while in service with the US Army's Pancho Villa Expedition into Mexico.[8] One notable instance was in May when the 6th Infantry received a reported sighting of Julio Cardenas, one of Villa's most trusted subordinates. Lt. George S. Patton led ten soldiers and two civilian guides in three Dodge Model 30 touring cars to conduct a raid at a ranch house in San Miguelito, Sonora. During the ensuing firefight the party killed three men, of whom one was identified as Cardenas. Patton's men tied the bodies to the hoods of the Dodges, returning to headquarters in Dublán and an excited reception from US newspapermen.

Death of the brothers

Dodge Brothers cars continued to rank second place in American sales in 1920. But that year, tragedy struck as John Dodge was felled by pneumonia in January. His brother Horace then died of cirrhosis in December of the same year (reportedly out of grief at the loss of his brother, with whom he was very close). The Dodge Brothers company fell into the hands of the brothers' widows, who promoted long-time employee Frederick Haynes to the company presidency. During this time, the Model 30 was evolved to become the new Series 116 (though it retained the same basic construction and engineering features).

1927 Dodge Brothers Series 124 sedan

Dodge Brothers emerged as a leading builder of light trucks. They also entered into an agreement whereby they marketed trucks built by Graham Brothers of Evansville, Indiana. The three Graham brothers would later produce Graham-Paige and Graham automobiles.

Stagnation in development was becoming apparent, however, and the public responded by dropping Dodge Brothers to fifth place in the industry by 1925. That year, the Dodge Brothers company was sold by the widows to the well-known investment group Dillon, Read & Co. for no less than US$146 million (at the time, the largest cash transaction in history).

Dillon, Read & Co. offered non-voting stock on the market in the new Dodge Brothers, Inc., firm, and along with the sale of bonds was able to raise $160 million, reaping a $14 million (net) profit. All voting stock was retained by Dillon, Read. Frederick Haynes remained as company head until E.G. Wilmer was named board chairman in November, 1926. Wilmer was a banker with no auto experience and Haynes remained as president. Changes to the car, save for superficial things like trim levels and colors, remained minimal until 1927, when the new Senior six-cylinder line was introduced. The former four-cylinder line was kept on, but renamed the Fast Four line until it was dropped in favor of two lighter six-cylinder models (the Standard Six and Victory Six) for 1928.

On October 1, 1925, Dodge Brothers, Inc., acquired a 51% interest in Graham Brothers, Inc., for $13 million and the remaining 49% on May 1, 1926. The three Graham brothers, Robert, Joseph and Ray, assumed management positions in Dodge Brothers before departing early in 1927.

Despite all this, Dodge Brothers' sales had already dropped to seventh place in the industry by 1927, and Dillon, Read began looking for someone to take over the company on a more permanent basis.

Purchase from Dillon, Read

Enter Walter P. Chrysler, head of the recently founded (in 1924) Chrysler Corporation and former president of General Motors' successful Buick division. Chrysler had wanted to purchase Dodge Brothers two years earlier, and had in the meantime created his own DeSoto brand of cars to challenge Dodge Brothers' new entries in the medium-priced field.

When Chrysler called again in 1928, Dillon, Read was finally ready to talk. In a foreshadowing of much later acquisitions by his company, Chrysler wanted Dodge Brothers more for its name, its extensive dealer network and its factory than anything it was producing at the time. The big sale came about in July 1928, when Chrysler and Dodge engaged in an exchange of stock worth $170 million. Production of existing models continued, with minor changes here and there, through the end of 1928 and (in the case of the Senior) into 1929. The new Chrysler-designed models for 1930 dropped the "Brothers" name and were marketed as just Dodge.

Pre-war years

Dodge D11 Luxury Liner 4-Door Sedan 1939

To fit better in the Chrysler Corporation lineup, alongside low-priced Plymouth and medium-priced DeSoto, Dodge's lineup for early 1930 was trimmed down to a core group of two lines and thirteen models (from three lines and nineteen models just over a year previous). Prices started out just above DeSoto but were somewhat less than top-of-the-line Chrysler, in a small-scale recreation of General Motors' "step-up" marketing concept. (DeSoto and Dodge would swap places in the market for the 1933 model year, Dodge dropping down between Plymouth and DeSoto.)

For 1930, Dodge took another step up by adding a new eight-cylinder line to replace the existing Senior six-cylinder. This basic format of a dual line with Six and Eight models continued through 1933, and the cars were gradually streamlined and lengthened in step with prevailing trends of the day. The Dodge Eight was replaced by a larger Dodge DeLuxe Six for 1934 and which was dropped for 1935. A long-wheelbase edition of the remaining Six was added for 1936 and would remain a part of the lineup for many years.

The Dodge line, along with most of the corporation's output, was restyled in the so-called "Wind Stream" look for 1935. This was a mild form of streamlining, which saw sales jump remarkably over the previous year (even though Dodge as a whole still dropped to fifth place for the year after two years of holding down fourth).

Another major restyle arrived for the 25th anniversary 1939 models, which Dodge dubbed the Luxury Liner series. These were once again completely redesigned with new bodies for 1940, again in 1941, and a refreshing for 1942. However, just after the 1942 models were introduced, Japan's attack on Pearl Harbor forced the shutdown of Dodge's passenger car assembly lines in favor of war production in February 1942.

Dodge

World War II

Chrysler was prolific in its production of war material from 1942 to 1945, and Dodge in particular was well-known to both average citizens and thankful soldiers for their tough military-spec truck models and ambulances like the WC54. Starting with the hastily converted VC series and evolving into the celebrated WC series, Dodge built a strong reputation for itself that readily carried over into civilian models after the war.

Post-war years

Civilian production at Dodge was restarted by late 1945, in time for the 1946 model year. The "seller's market" of the early postwar years, brought on by the lack of any new cars throughout the war, meant that every automaker found it easy to sell vehicles regardless of any drawbacks they might have. Like almost every other automaker, Dodge sold lightly facelifted revisions of its 1942 design through the 1948 season. As before, these were a single series of six-cylinder models with two trim levels (basic Deluxe or plusher Custom).

Dodge Coronet 1955

Styling was not initially Dodge's strong point during this period, though that began to change by 1953 under the direction of corporate design chief Virgil Exner. At the same time, Dodge also introduced its first V8 engine — the Red Ram Hemi, a smaller version of the original design of the famed Hemi. The new 1953 bodies were smaller and based on the Plymouth. For 1954, sales dropped, the stubby styling not going over well with the public.

New corporate "Forward Look" styling for 1955 began a new era for Dodge. With steadily upgraded styling and ever-stronger engines every year through 1960, Dodge found a ready market for its products as America discovered the joys of freeway travel. This situation improved when Dodge introduced a new line of Dodges called the Dart to do battle against Ford, Chevrolet and Plymouth. The result was that Dodge sales in the middle price class collapsed, and the Polara was dropped at the end of 1961.

1958 Dodge Coronet Lancer hardtop coupe

Dodge entered the compact car field for 1961 with their new Lancer sedan (a variation on Plymouth's Valiant). Though it was not initially successful, the Dart range that came after it in 1963 would prove to be one of the division's top sellers for many years.

Chrysler did make an ill-advised move to downsize the Dodge and Plymouth full-size lines for 1962, which resulted in a loss of sales. However, they turned this around in 1965 by turning those former full-sizes into "new" mid-size models; Dodge revived the Coronet nameplate in this way and later added a sporty fastback version called the → Charger that became both a sales leader and a winner on the NASCAR circuit.

Full-size models evolved gradually during this time. After Dodge dealers complained about not having a true full-size car in the fall of 1961, the Custom 880 was hurried into production. The Custom 880 used the 1962 Chrysler Newport body with the 1961 Dodge front end and interior. The 880 continued into 1965, the year a completely new full-size body was put into production, the Polara entered the medium price class and the Monaco was added as the top series. The Polara and Monaco were changed mostly in appearance for the next ten years or so. Unique "fuselage" styling was employed for 1969 through 1973 and then was toned down again for the 1974 to 1977 models.

Dodge is well-known today for being a player in the → muscle car market of the late 1960s and early 1970s. Along with the Charger, models like the Coronet R/T and Super Bee were popular with buyers seeking performance. The pinnacle of this effort was the introduction of the Challenger sports coupe and convertible (Dodge's entry into the "pony car" class) in 1970, which offered everything from mild economy engines up to the wild race-ready Hemi V8 in the same package.

1966 Dodge Coronet 440 sedan

In an effort to reach every segment of the market, Dodge even reached a hand across the Pacific to its partner, Mitsubishi Motors, and marketed their subcompact as the Colt to compete with the AMC Gremlin, Chevrolet Vega, and Ford Pinto. Chrysler would over the years come to rely heavily on their relationship with Mitsubishi.

Times of crisis

Everything changed at Dodge (and Chrysler as a whole) when the 1973 oil crisis hit the United States. Save for the Colt and certain models of the Dart, Dodge's lineup was quickly seen as extremely inefficient. In fairness, this was true of most American automakers at the time, but Chrysler was also not in the best financial shape to do anything about it. Consequently, while General Motors and Ford were quick to begin downsizing their largest cars, Chrysler (and Dodge) moved more slowly out of necessity.

1977 Dodge Diplomat sedan

At the very least, Chrysler was able to use some of its other resources. Borrowing the recently-introduced Chrysler Horizon from their European division, Dodge was able to get its new Omni subcompact on the market fairly quickly. At the same time, they increased the number of models imported from Mitsubishi: first came a smaller Colt (based on Mitsubishi's Mitsubishi Lancer line), then a revival of the Challenger (though with nothing more than a four-cylinder under the hood, rather than the booming V8s of yore).

Bigger Dodges, though, remained rooted in old habits. The Dart was replaced by a new Aspen for 1976, and Coronet and Charger were effectively replaced by the Diplomat for 1977, which was actually a fancier Aspen. Meanwhile, the huge Monaco (Royal Monaco beginning in 1977 when the mid-sized Coronet was renamed "Monaco") models hung around through 1977, losing sales every year, until finally being replaced by the St. Regis for 1979 following a one-year absence from the big car market. In a reversal of what happened for 1965, the St. Regis was an upsized Coronet. Buyers, understandably, were confused and chose to shop the competition rather than figure out what was going on at Dodge.

Everything came to a head in 1979 when Chrysler's new chairman, Lee Iacocca, requested and received federal loan guarantees from the United States Congress in an effort to save the company from having to file bankruptcy. With bailout money in hand, Chrysler quickly set to work on new models that would leave the past behind.

K-Cars and minivans

1981–82 Aries Special Edition

The first fruit of Chrysler's crash development program was the "K-Car," the Dodge version of which was the Dodge Aries. This basic and durable front-wheel drive platform spawned a whole range of new models at Dodge during the 1980s, including the groundbreaking Dodge Caravan. The Caravan not only helped save Chrysler as a serious high-volume American automaker, but also spawned an entirely new market segment that remains popular today: the minivan.

Through the late 1980s and 1990s, Dodge's designation as the sporty-car division was backed by a succession of high-performance and/or aggressively-styled models including the Daytona, mid-sized 600 and several versions of the Lancer. The Dodge Spirit sedan was well received in numerous markets worldwide. The Omni remained in the line through 1990. Dodge-branded Mitsubishi vehicles were phased out by 1993 with the exception of the Dodge Stealth running through 1996, though Mitsubishi-made engines and electrical components were still widely used in American domestic Chrysler products. In 1992, Dodge moved their performance orientation forward substantially with the Viper, which featured an aluminum V10 engine and composite sports roadster body. This was the first step in what was marketed as "The New Dodge." Step two was the new Intrepid sedan, totally different from its boxy Dynasty predecessor.

1991 Dodge Spirit R/T

The Intrepid used what Chrysler called "cab forward" styling, with the wheels pushed out to the corners of the chassis for maximum passenger space. They followed up on this idea in a smaller scale with the Stratus and Neon, both introduced for 1995. The Neon in particular was a hit, buoyed by a clever marketing campaign and good performance.

The modern era

Cab Forward Design on a 1996 Dodge Stratus

DaimlerChrysler

Chrysler Corporation was sold to Daimler-Benz AG in 1998 to form DaimlerChrysler. Rationalizing Chrysler's broad lineup was a priority, and Dodge's sister brand Plymouth was withdrawn from the market. With this move, Dodge became DaimlerChrysler's low-price division as well as its performance division.

2006 Dodge Charger SRT8 sedan

The Intrepid, Stratus, and Neon updates of the 1998 to 2000 timeframe were largely complete before Daimler's presence, and Dodge's first experience of any synergy with the German side of the company was the 2005 Magnum station wagon, introduced as a replacement for the Intrepid. Featuring Chrysler's first mainstream rear-wheel drive platform since the 1980s and a revival of the Hemi V8 engine, it was a modest success. The Charger was launched in 2006 on the same platform.

Further synergies were explored in the form of an extensive platform-sharing arrangement with Mitsubishi, which spawned the Caliber subcompact as a replacement for the Neon and the Avenger sedan. The rear-drive chassis was then used in early 2008 to build a new Challenger, with styling reminiscent of the original 1970 Challenger.

In Spring 2007, DaimlerChrysler reached an agreement with Cerberus Capital Management to sell off its Chrysler Group subsidiary, of which the Dodge division was a part. On June 10, 2009, Italian automaker Fiat formed a partnership with Chrysler in which a "New Chrysler" was formed and was given the name Chrysler Group LLC, which Dodge remains a part of.

In response to very high motor fuel prices in Spring 2008, Dodge initiated a purchase incentive guaranteeing the buyer of a new Dodge would have to pay no more than $2.99 per gallon of gasoline for three years. Shortly after the promotion began, the average price of gasoline dropped well below $2.99 per gallon.

Dodge Trucks

Over the years, Dodge has become at least as well-known for its many truck models as for its prodigious passenger car output.

Pickups and medium to heavy trucks

Ever since the beginning of its history in 1914, Dodge has offered light truck models to interested buyers. For the first few years, these were based largely on the existing passenger cars, but eventually gained their own chassis and body designs as the market matured. Light- and medium-duty models were offered first, then a heavy-duty range was added during the 1930s and 1940s.

Following World War II and the successful application of four-wheel drive to the truck line, Dodge introduced a civilian version that it called the Power Wagon. At first based almost exactly on the military-type design, variants of the standard truck line were eventually given 4WD and the same "Power Wagon" name.

Dodge was among the first to introduce car-like features to its trucks, adding the plush Adventurer package during the 1960s and offering sedan-like space in its Club Cab bodies of the 1970s. Declining sales and increased competition during the 1970s eventually forced the company to drop its medium- and heavy-duty models, an arena the company has only recently begun to reenter.

Dodge introduced what they called the "Adult Toys" line to boost its truck sales in the late 1970s, starting off with the limited edition Lil' Red Express pickup (featuring visible big rig-style exhaust stacks). Later came the more widely available Warlock. Other "Adult Toys" from Dodge included the Macho Power Wagon and Street Van.

As part of a general decline in the commercial vehicle field during the 1970s, Dodge eliminated their LCF Series heavy-duty trucks in 1975, along with the Bighorn and medium-duty D-Series trucks, and affiliated S Series school buses were dropped in 1978. On the other hand, Dodge produced several thousand pickups for the United States Military under the CUCV program from the late 1970s into the early 1980s.

1989 Dodge Ram pickup

Continuing financial problems meant that even Dodge's light-duty models – renamed as the Ram Pickup line for 1981 – were carried over with the most minimal of updates until 1993. But two things helped to revitalize Dodge's fortunes during this time. First was their introduction of Cummins' powerful and reliable B Series turbo-diesel engine as an option for 1989. This innovation raised Dodge's profile among serious truck buyers who needed big power for towing or large loads. A compact Dakota pickup, which later offered a class-exclusive V8 engine, was also an attractive draw.

Dodge introduced the Ram's all-new "big-rig" styling treatment for 1994. Besides its instantly polarizing looks, exposure was also gained by usage of the new truck on the hit TV show *Walker, Texas Ranger* starring Chuck Norris. The new Ram also featured a totally new interior with a console box big enough to hold a laptop computer, or ventilation and radio controls that were designed to be easily used even with gloves on. A V10 engine derived from that used in the Viper sports car was also new, and the previously offered Cummins turbo-diesel remained available. The smaller Dakota was redesigned in the same vein for 1997, thus giving Dodge trucks a definitive "face" that set them apart from the competition.

The Ram was redesigned again for 2003 (the Dakota in 2002), basically as an evolution of the original but now featuring the revival of Chrysler's legendary Hemi V8 engine. New medium-duty chassis-cab models were introduced for 2007 (with standard Cummins turbo-diesel power), as a way of gradually getting Dodge back in the business truck market again.

For a time during the 1980s, Dodge also imported a line of small pickups from Mitsubishi. Known as the D50 or (later) the Ram 50, they were carried on as a stopgap until the Dakota's sales eventually made the imported trucks irrelevant. (Ironically, Mitsubishi has more recently purchased Dakota pickups from Dodge and restyled them into their own Raider line for sale in North America.)

Vans

Dodge had offered panel delivery models for many years since its founding, but their first purpose-built van model arrived for 1964 with the compact A Series. Based on the Dodge Dart platform and using its proven six-cylinder or V8 engines, the A-series was a strong competitor for both its domestic rivals (from Ford and Chevrolet/GMC) and the diminutive Volkswagen Transporter line.

As the market evolved, however, Dodge realized that a bigger and stronger van line would be needed in the future. Thus the B Series, introduced for 1971, offered both car-like comfort in its Sportsman passenger line or expansive room for gear and materials in its Tradesman cargo line. A chassis-cab version was also offered, for use with bigger cargo boxes or flatbeds.

Like the trucks, though, Chrysler's dire financial straits of the late 1970s precluded any major updates for the vans for many years. Rebadged as the Ram Van and Ram Wagon for 1981, this venerable design carried on with little more than cosmetic updates all the way to 2003.

The DaimlerChrysler merger of 1999 made it possible for Dodge to explore new ideas; hence the European-styled Mercedes-Benz Sprinter line of vans was brought over and given a Dodge styling treatment. Redesigned for 2006 as a 2007 model, the economical diesel-powered Sprinters have become very popular for city usage among delivery companies like FedEx and UPS in recent years.

Dodge also offered a cargo version of its best-selling Caravan for many years, at first calling it the Mini Ram Van (a name originally applied to short-wheelbase B-Series Ram Vans) and later dubbing it the Caravan C/V (for "Cargo

Van").

Sport utility vehicles

Dodge's first experiments with anything like a sport utility vehicle were seen in the late 1950s with a windowed version of their standard panel truck known as the Town Wagon. These were built in the same style through the mid-1960s.

But the division didn't enter the SUV arena in earnest until 1974, with the purpose-built Ramcharger. Offering the then-popular open body style and Dodge's powerful V8 engines, the Ramcharger was a strong competitor for trucks like the Ford Bronco, Chevrolet Blazer and International Harvester Scout II.

Once again, though, Dodge was left with outdated products during the 1980s as the market evolved. The Ramcharger hung on through 1993 with only minor updates, but was not replaced along with the rest of the truck line for 1994.

Instead, Dodge tried something new in 1998. Using the mid-sized Dakota pickup's chassis as a base, they built the four-door Durango SUV with seating for seven people and created a new niche. Sized between smaller SUVs (like the Chevrolet Blazer and Ford Explorer) and larger models (like the Chevrolet Tahoe and Ford Expedition), Durango was both a bit more and bit less of everything. The redesigned version for 2004 grew a little bit in every dimension, becoming a full-size SUV (and was thus somewhat less efficient), but was still sized between most of its competitors on either side of the aisle.

Dodge also imported a version of Mitsubishi's popular Montero (Pajero in Japan) as the Raider from 1987 to 1989.

International markets

Dodge vehicles are now available in many countries throughout the world.

Asia

Dodge entered the Japanese market in mid-2007, and re-entered the Chinese market in late 2007. Soueast Motors of China assembles the Caravan for the Chinese market. Dodge had already been marketing its vehicles in South Korea since 2004, starting with the Dakota.

Dodge vehicles have been sold in the Middle East for a considerably longer period of time.

Australia

Dodge recently re-entered the Australian market in 2006 after a 30-year absence. Dodge Australia plans to release a new model every six months for the next three years, amid plans to re-ignite the brand's interest Down Under. The first of such models is the Dodge Caliber, which was well received at the recent 2006 Melbourne International Motor Show. The second model to be introduced was the Nitro, and the Avenger has also recently joined the lineup.

Brazil

In Brazil, Dodge cars have been successful with the models Dakota and Ram, recently the only available model was the Ram 2500, but the model portfolio is being expanded, starting with the Journey crossover for the 2009 model year.

Canada

In Canada, the Dodge lineup of cars started down the road to elimination along with the Plymouth line when in 1988 the Dodge Dynasty was sold in Canada as the Chrysler Dynasty and sold at both Plymouth and Dodge dealers. Similarly, the new Dodge Intrepid, the Dynasty's replacement, was sold as the Chrysler Intrepid.

For 2000, the new Neon became the Chrysler Neon. The Chrysler Cirrus and Mitsubishi-built Dodge Avenger were dropped. Dodge trucks, which have been sold at Canadian Plymouth dealers since 1973, continued without change. All Plymouth-Chrysler and Dodge-Chrysler dealers became Chrysler-Dodge-Jeep dealers.

The cheapening of the Chrysler name did not go well in Canada, especially as the nameplate had been pushed as a luxury line since the 1930's, and for 2003 the revamped Neon appeared in Canada as the Dodge SX 2.0. Since then all new Dodge models have been sold in Canada under the Dodge name.

Europe

Following Chrysler's takeover of the British Rootes Group, Simca of France, and Barreiros of Spain, and the resultant establishment of Chrysler Europe in the late 1960s, the Dodge brand was used on light commercial vehicles, most of which were previously branded Commer or Karrier, on pickup and van versions of the Simca 1100, on the Spanish Dodge Dart, and on heavy trucks built in Spain. The most common of these was the Dodge 50 series, widely used by utility companies and the military, but rarely seen outside the UK, and the Spanish-built heavy-duty 300 series available as 4x2, 6x4, 8x2, and 8x4 rigids, as well as 4x2 semi-trailer tractors. All of these were also sold in selected export markets badged either as Fargo or De Soto.

Following Chrysler Europe's collapse in 1977, and the sale of their assets to Peugeot, the Chrysler/Dodge British and Spanish factories were quickly passed on to Renault Véhicules Industriels, who gradually re-branded the range of vans and trucks as Renaults through the 1980s. They would eventually drop these products altogether and used the plants to produce engines (in the UK) and "real" Renault truck models in Spain. Dodge vehicles would not return to the UK until the introduction of the Dodge Neon SRT-4, branded as a Chrysler Neon, in the mid 2000s.

The Dodge marque was reintroduced to Europe on a broad scale in 2006. Currently, the Dodge lineup in Europe consists of the Caliber, Avenger, Viper SRT-10, Nitro and Dodge Journey (2008).

Mexico

In Mexico, the Hyundai Accent, Hyundai Atos, and Hyundai H100 are branded as "Dodge" or "Verna by Dodge", "Atos by Dodge" and "Dodge H100" respectively, and sold at Chrysler/Dodge dealers.

Marketing

Dodge Quest Project Committee designed a bilingual (English/Japanese) RPG game called Dodge Quest where player plays a man named Coronet.[9][10]

Logos

- **Star** The original Dodge logo was round, with two interlocking triangles forming a six-pointed star in the middle; an interlocked "DB" was at the center of the star, and the words "Dodge Brothers Motor Vehicles" encircled the outside edge. Although the "Brothers" was dropped from the name for trucks in 1929 and cars in 1930, the DB star remained in the cars until the 1939 models were introduced.
- **Ram** For 1932 Dodge cars adopted a leaping ram as the car's hood ornament. Starting with the 1940 models the leaping ram became more streamlined and by 1951 only the head, complete with curving horns, remained. The 1954 model cars were the last to use the ram's head before the rebirth in the 1980's. Dodge trucks adopted the ram as the hood ornament for the 1940 model year with the 1950 models as the last.

- **Crest** For 1941 Dodge introduced a crest, supposedly the Dodge family crest. The design had four horizontal bars broken in the middle by one vertical bar with an "O" in the centre. A knight's head appeared at the top of the emblem. Although the head would be dropped for 1955, the emblem would survive through 1957 and reappear on the 1976 Aspen. The crest would be used through to 1981 on its second time around, being replaced by the pentastar for 1982. The knight's head without the crest would be used for 1959.
- **Forward Look** Virgil Exner's radical "Forward Look" redesign of Chrysler Corporation's vehicles for the 1955 model year was emphasized by the adoption of a logo by the same name, applied to all Chrysler Corporation vehicles. The Forward Look logo consisted of two overlapped boomerang shapes, suggesting space age rocket-propelled motion. This logo was incorporated into Dodge advertising, decorative trim, ignition and door key heads, and accessories through September 1962. *See also:* Forward Look
- **Fratzog** Dodge's logo from September 1962 though 1976 was a fractured deltoid composed of three arrowhead shapes forming a 3-pointed star. The logo first appeared on the 1962 Polara 500 and the mid-year 1962 Custom 880. One of its designers came up with the meaningless name **Fratzog** for the logo, which ultimately stuck.[11] [12] As the Dodge Division's logo, Fratzog was incorporated in various badges and emblems on Dodge vehicles. It was also integrated into the design of such parts as steering wheel center hubs and road wheel covers.
- **Pentastar** From 1982 to 1992 Dodge used Chrysler's Pentastar logo on its cars and trucks to replace the Dodge crest, although it had been used for corporate recognition since March, 1963. In advertisements and on dealer signage, Dodge's Pentastar was red, while Chrysler-Plymouth's was blue.
- **Ram's head** Dodge reintroduced the ram's head hood ornament on the new 1973 Dodge Bighorn heavy duty tractor units. Gradually the ram's head began appearing on the pickup trucks as Dodge began to refer to their trucks as Ram. The present iteration of the Ram's-head logo appeared in 1993, standardizing on that logo in 1996 for all vehicles except the Viper.
- **New logo** In 2010, with the separation of the Ram brand, the new Dodge logo will remove the ram's head and just read "DODGE" on a black background with a touch of red. [13]

Dodge Brothers emblem ca. 1910, removed from the gate of the Dodge Main plant before its 1981 demolition

Dodge Brothers logo used from 1914-27 (seen here on a modern belt buckle)

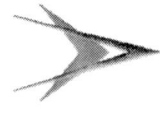

Forward Look logo

Fratzog

Red Chrysler Pentastar logo, used 1976-1992

Models

As of 2008, Dodge's model range in North America consists of the Caliber, Avenger, Journey, → Charger, Grand Caravan, Challenger, and Viper passenger cars, the Dakota and Ram pickup trucks, the Nitro and the Durango SUVs, and the Sprinter van.

See also

- Chrysler LLC
- List of Dodge automobiles for all production cars and trucks
- List of Dodge concept vehicles.
- Plymouth (automobile)
- DeSoto (automobile)
- List of automobile manufacturers
- Rootes Group for the historic Dodge of the UK.

References

- Brinkley, Douglas. (2004) *Wheels for the World: Henry Ford, his Company, and a Century of Progress, 1903–2003*. ISBN 0142004391.
- Burness, Tad. (2001) *Ultimate Truck & Van Spotter's Guide 1925–1990*. ISBN 0-87341-969-3.
- Gunnell, John, Editor (1987). *The Standard Catalog of American Cars 1946–1975*. Kraus Publications. ISBN 0-87341-096-3.
- Gunnell, John A., ed. (1993) *Standard Catalog of American Light-Duty Trucks, Second Edition*. ISBN 0-87341-238-9.
- Lenzke, James T., ed. (2000) *Standard Catalog of Chrysler 1914-2000*. ISBN 0-87341-882-4.
- Ruiz, Marco. (1986) *Japanese Car*. ISBN 0-517-61777-3.
- Vlasic, Bill and Stertz, Bradley A. (2000) *Taken for a Ride: How Daimler-Benz Drove Off with Chrysler*. ISBN 0-688-17305-5.

External links

- Dodge USA [1]
- Dodge Canada [14]
- Dodge Mexico [15]
- Dodge Italy [16]
- Chrysler LLC Worldwide [17]
- Allpar Mopar Vehicles [18]
- FleetData: *History of Dodge in the UK* [19]—website of the Road Transport Fleet Data Society
- ww2dodge.com [20]—WW II Dodge Truck History: site for military Dodge's produced 1939–1945
- Old Dodges.com [21]—Site devoted to Dodge Medium and Heavy-Duty Trucks of the 1960s and 1970s, primarily focusing on the Dodge Bighorn Trucks (1973-1975).

References

[1] http://www.dodge.com/
[2] http://www.international.chryslercorp.com/dindex.html
[3] money (2009-10-26). " Latest Dispatches- MSN Money (http://articles.moneycentral.msn.com/Investing/Dispatch/default. aspx?feat=1337558)". Articles.moneycentral.msn.com. . Retrieved 2009-11-19.
[4] Linebaugh, Kate (2009-10-27). " Fiat Models to Drive Chrysler - WSJ.com (http://online.wsj.com/article/SB125659536562909009. html)". Online.wsj.com. . Retrieved 2009-11-19.
[5] Wise, David Burgess. "Dodge: ", in *World of Automobiles* (London: Orbis Publishing Ltd., 1974), Volume 5, p.552.
[6] Wise, p.551.
[7] Wise, p.552.
[8] '"The Mexican Revolution 1910-20" by P Jowett & A de Quesada, Osprey Publishing Ltd., p. 25, ISBN 1-84176-989-4'
[9] DQ page (http://dodge-quest.com/)
[10] Dodge Quest + (http://xorsyst.com/games/dodge-quest/)
[11] 1962-1964 Dodge 880 article (http://auto.howstuffworks.com/1962-1964-dodge-8802.htm)
[12] Mike Sealey's history of Chrysler's logos (http://www.allpar.com/history/logos.html)
[13] Riches, Erin (2009-11-04). " Chrysler-Fiat Press Conference: Only 500 More Vipers Will Be Built (http://blogs.insideline.com/ straightline/2009/11/chrysler-fiat-press-conference-only-500-more-vipers-will-be-built.html)". Blogs.insideline.com. . Retrieved 2009-11-19.
[14] http://www.dodge.ca
[15] http://www.dodge.com.mx
[16] http://www.dodge.it/
[17] http://www.international.chryslercorp.com
[18] http://www.allpar.com/
[19] http://fleetdata.co.uk/dodge.html
[20] http://ww2dodge.com/
[21] http://www.olddodges.com/

Chrysler B platform

The **Chrysler B platform** was the basis for rear-wheel drive → Chrysler cars from 1962 through 1979. All of the B-body cars in a given model year for either make were built upon the same chassis. However, the outward design differed between makes.

The Plymouth B-body series ultimately comprised four cars with nearly identical outward appearances (differing only in trim package, drive train and accessories). These were the Belvedere, Satellite, GTX and Road Runner. The Superbird was a Road Runner with an extended nose and a high-mounted rear wing. It was the only Plymouth B-body that looked essentially different from the others.

There was more diversity in the outward appearance of the Dodge B-body series. The Dodge models based on the B-body were the Coronet, Super Bee and the → Charger. The Charger Daytona was a Charger with an extended nose and high-mounted rear wing.

Cars using the rear wheel drive B platform include:

- 1962 Dodge Dart
- 1962-1964 Dodge Polara
- 1962-1964 Plymouth Fury
- 1962-1964 Plymouth Savoy
- 1962-1970 Plymouth Belvedere
- 1963-1964 Dodge 220 (Canadian)
- 1963-1964 Dodge 330
- 1963-1964 Dodge 440
- 1965-1974 Plymouth Satellite
- 1965-1976 Dodge Coronet

- 1966-1978 → Dodge Charger
- 1967-1971 Plymouth GTX
- 1968-1975 Plymouth Road Runner
- 1975-1978 Plymouth Fury
- 1975-1979 → Chrysler Cordoba
- 1977-1978 Dodge Monaco
- 1978-1979 Dodge Magnum
- 1979 Chrysler 300

Five different wheelbases were available:

- 116 in
  - 1962 Dodge Dart
  - 1962-1964 Dodge Polara
  - 1962-1966 Plymouth wagons
  - 1962-1970 Plymouths (except wagons)
  - 1963-1964 Dodge 220/330/440
- 115 in
  - 1971-1979 2-door models
  - 1975-1979 Chrysler Cordoba
- 117 in
  - 1965-1970 → Dodges
  - 1967-1974 Plymouth wagons
  - 1971-1974 Plymouth 4 doors
- 117.5 in
  - 1975-1978 Plymouth and → Dodge 4 doors and wagons
- 118 in
  - 1971-1974 → Dodge

1988-1992

From 1988 to 1992, the B-body name was used again for the midsize front wheel drive Eagle Premier sedan, which was originally designed by and was slated to be built by American Motors with Renault until Chrysler's buyout of that company in March 1987. The Premier was later joined by the similar Dodge Monaco for 1990.

Models

- 1988-1992 Eagle Premier
- 1990-1992 Dodge Monaco

See also

- Chrysler platforms

External links

- B Body Mopar Forum [1]
- Dodge Charger [1]
- Plymouth Road Runner [2]
- Plymouth Superbird [3]
- 1969 Roadrunner Forum [4]
- B-bodies [5] at Lee Herman's Mopar Page (three pages).

References

[1] http://www.forbbodiesonly.com
[2] http://www.stockmopar.com/plymouth-road-runner.html
[3] http://www.stockmopar.com/plymouth-superbird.html
[4] http://www.69roadrunner.net
[5] http://www.lhmopars.com/Nats-BBody1.htm

Chrysler Cordoba

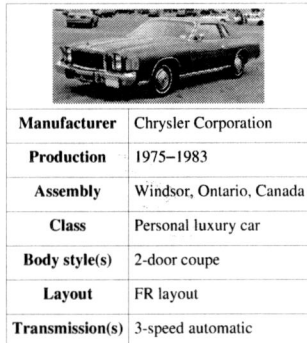

Manufacturer	Chrysler Corporation
Production	1975–1983
Assembly	Windsor, Ontario, Canada
Class	Personal luxury car
Body style(s)	2-door coupe
Layout	FR layout
Transmission(s)	3-speed automatic

The **Chrysler Cordoba** was an intermediate personal luxury coupe sold by Chrysler Corporation in North America from 1975 to 1983. It was the company's first model produced specifically for the personal luxury market and the first Chrysler-branded vehicle that was less than full-size.

History

In the early 1960s, when other upmarket brands were expanding into smaller cars with such models as the Mercury Comet and Buick Skylark, Chrysler very publicly declared that there would "never" be a smaller Chrysler. Historians of the marque noted later that "never" on the Chrysler timeline had equaled not quite fifteen years. The Cordoba was one of Chrysler's few genuine hits of the 1970s. At a time when Chrysler was teetering on bankruptcy, demand actually exceeded supply for its first couple of years, with production of over 150,000 units annually. Half of Chrysler division production during this period (and occasionally more) was composed of Cordobas. Cordobas were built in Windsor, Ontario.

Although Córdoba is the name of a city in Spain, the car's emblem was actually a stylized version of the Argentine cordoba coin. Either way, the implication was Hispanic, and this theme was carried out with somewhat baroque trim inside and by using movie star Ricardo Montalban as the car's advertising spokesman. Notable was his eloquent praise of its "soft Corinthian leather" interior[1] and his Americanized stress on the second syllable of the car's name.

First generation (1975-1979)

Production	1975–1979
Platform	→ B-body
Engine(s)	318 cu in (5.2 L) *LA* V8 360 cu in (5.9 L) *LA* V8 400 cu in (6.6 L) *B*
Transmission(s)	3-speed *A727* automatic
Wheelbase	115 in (2921 mm)
Length	215.3 in (5469 mm)
Width	77.1 in (1958 mm)
Height	52.6 in (1336 mm)
Related	Chrysler 300 → Dodge Charger Dodge Magnum

The Cordoba was first introduced for 1975, as an upscale personal luxury car. At the time the personal luxury market was large and growing, with the Chevrolet Monte Carlo and Pontiac Grand Prix selling over 300,000 units each annually. The car carried the Chrysler name, then still associated exclusively with large luxury models like the Imperial. It was, however, priced to compete with rivals such as the Monte Carlo, Ford Elite, and Oldsmobile Cutlass Supreme. The Cordoba was originally intended to be a Plymouth (the names Mirada, Premier, and Grand Era were associated with the project), but losses from the newly introduced full-size C-body models in 1974 (at the onset of the energy crisis) encouraged Chrysler executives to seek higher profits by marketing the model as a Chrysler, a name with a more upscale appeal. The car was an unforseen sucess, with over 150,000 examples sold in 1975, a sales year that was otherwise dismal for the company. For 1976 sales increased slightly to 165,000. The midly tweaked 1977 version also sold well, with just under 140,000 cars sold. The success of this strategy is well illustrated by the fact that its similar and somewhat cheaper corporate cousin, the → Dodge Charger SE, only sold about a quarter as well during the same model years.

1978-1979 Cordoba, with rectangular headlights

The original design endured with only very small changes for three years before a variety of factors contributed to a decline in sales. For 1978, there was a modest restyling with the then de rigueur rectangular headlights in a stacked configuration that had the unfortunate effect of making the Cordoba look much like the 1976 to 1977 Monte Carlo from the front. A Chrysler designer, Jeffrey Godshall, wrote in his article on the Cordoba in *Collectible Automobile* magazine that this restyling was viewed as "somewhat tacky" and eliminated much of the visual appeal that the 1975 to 1977 Cordobas had been known for. The restyle also made the car appear heavier than its 1975-77 predecessor.

At the same time, Chrysler's financial position and quality reputation was in steady decline, and rising gas prices and tightening fuel economy standards made the Cordoba's nearly 4000 lb (1814 kg) weight with 360 cu in (5.9 L) or 400 cu in (6.6 L) V8 engines obsolete. In its final year, 1979, however, high performance made a return as the original Cordoba provided the platform for a one-year-only revival of the Chrysler 300 name.

Second generation (1980-1983)

Production	1980–1983
Platform	J-body
Engine(s)	3.7 L *Slant* 6 I6 5.2 L *LA* V8 5.9 L *LA* V8
Wheelbase	112.7 in (2863 mm)
Length	210.1 in (5337 mm) LS: 209.6 in (5324 mm)
Width	72.7 in (1847 mm)
Height	53.2 in (1351 mm)
Related	Dodge Mirada Imperial (1981-1983)

The Cordoba was downsized for the 1980 model year. The new smaller model used the J-platform, which dated back to the 1976 Plymouth Volaré and was twinned up with the newly-named but very similar Dodge Mirada. Chrysler also revived the Imperial for 1981 as a third variant of the J-platform. The Cordoba and Mirada now had a standard six-cylinder engine (the famous 225 Slant Six), which, while very reliable, did not seem to be suitable power for these slightly upmarket coupes. The much-detuned >318 cu in (5.2 L) V8 was an option (standard on the Imperial), along with (for 1980 only) the 360 cu in (5.9 L) V8, which was in its final year in Chrysler's cars.

The second-generation Cordoba's styling did not attract the praise of the original, and sales were off substantially. It is true that downsizing was tough on personal luxury models generally; both the Monte Carlo in 1978 and the 1980 Ford Thunderbird shrank in size and sales simultaneously. However, those models eventually recovered as their makers moved to correct their cars' flaws, while the smaller Cordoba never did. Chrysler was increasingly concentrating on its compact, front wheel drive models with modern four and six-cylinder engines, and management stopped producing the Cordoba in 1983. Total sales of the second generation cars was just under 100,000 units.

NASCAR

Both the first generation and second generation Cordoba's made appearances in NASCAR. Ed Negre campaigned one occasionally in 1979-80 seasons, and Buddy Arrington ran a second generation car in the 1982-84 seasons, alternating with Dodge Miradas and Chrysler Imperials. The Cordoba was no more aerodynamic than the other Mopars and never finished higher that 15th in any race entered.

Today

Today, the model has a fairly loyal owner base and some models are considered collectible. The very early production 1975s, particularly with the optional four-barrel carburetor, and the Cordoba-based 300 of 1979 are the most valuable. The second generation Cordoba has attracted little interest in the collector market so far except for the rare few LS models, which sported a stylish aerodynamic nosecone with "crosshair" grille. Other features of this model were vinyl top delete and monotone coloring.

External links

- Cordoba page at Allpar.com [2]

References

[1] Frum, David (2000). *How We Got Here: The '70s*. New York, New York: Basic Books. p. 25. ISBN 0465041957.
[2] http://www.allpar.com/model/cordoba.html

Chrysler

★ CHRYSLER	
Type	Limited liability company
Predecessor	Chrysler LLC
Founded	June 6, 1925
Founder(s)	Walter Chrysler
Headquarters	Auburn Hills, Michigan, U.S.
Number of locations	List of Chrysler factories
Key people	C. Robert Kidder (Chairman)[1] Sergio Marchionne (CEO)[2]
Industry	Automotive
Products	→ Automobiles
Owner(s)	UAW VEBA (67.69%) Fiat S.p.A. (20%) U.S. Government (9.85%) Government of Canada (2.46%)[3]
Employees	58,000 (2008)
Divisions	Chrysler → Dodge Jeep Ram Mopar Global Electric Motorcars(GEM)[4]
Subsidiaries	Chrysler Australia Chrysler Canada GEM
Website	Chryslergroupllc.com [5]

Chrysler Group, LLC is a U.S. automobile manufacturer headquartered in the Detroit suburb of Auburn Hills, Michigan. Chrysler was first organized as the Chrysler Corporation in 1925.[6] From 1998 to 2007, Chrysler and its subsidiaries were part of the German based DaimlerChrysler AG (now Daimler AG).[7] Prior to 1998, Chrysler Corporation traded under the "C" symbol on the New York Stock Exchange. Under DaimlerChrysler, the company was named "DaimlerChrysler Motors Company LLC", with its U.S. operations generally referred to as the "Chrysler Group". On May 14, 2007, DaimlerChrysler announced the sale of 80.1% of Chrysler Group to American private equity firm Cerberus Capital Management, L.P., although Daimler continued to hold a 19.9% stake. This was when the company took on the name, Chrysler LLC.[8] The deal was finalized on August 3, 2007.[9] On April 27, 2009, Daimler AG signed a binding agreement to give up its 19.9% remaining stake in Chrysler LLC to Cerberus Capital Management and pay as much as $600 million into the automaker's pension fund.[10]

On April 30, 2009 Chrysler LLC filed for Chapter 11 bankruptcy protection and announced a plan for a partnership with Italian automaker Fiat.[11] On June 1, Chrysler LLC stated they were selling some assets and operations to the newly formed company Chrysler Group LLC.[12] Fiat will hold a 20% stake in the new company, with an option to increase this to 35%, and eventually to 51% if it meets financial and developmental goals for the company.[13]

On June 10, 2009, the sale of most of Chrysler assets to "New Chrysler", formally known as Chrysler Group LLC was completed. The federal government financed the deal with US$6.6 billion in financing, paid to the "Old Chrysler", formally called Old Carco LLC.[14] The transfer does not include eight manufacturing locations, nor many parcels of real estate, nor equipment leases. Contracts with 789 U.S. auto dealerships, who are being dropped, were not transferred.[15] [16]

History

The company was founded by Walter P. Chrysler on June 6, 1925,[17] when the Maxwell Motor Company (est. 1904) was re-organized into the Chrysler Corporation.[18] [19]

Walter Chrysler had originally arrived at the ailing Maxwell-Chalmers company in the early 1920s, having been hired to take over and overhaul the company's troubled operations (just after a similar rescue job at the Willys car company).[20]

In late 1923 production of the Chalmers automobile was ended.[21]

Then in January 1924, Walter Chrysler launched the well-received Chrysler automobile. The Chrysler was a 6-cylinder automobile, designed to provide customers with an advanced, well-engineered car, but at a more affordable price than they might expect. (Elements of this car are traceable back to a prototype which had been under development at Willys at the time that Walter Chrysler was there).[22] The original 1924 Chrysler included a carburetor air filter, high compression engine, full pressure lubrication, and an oil filter, at a time when most autos came without these features.[23] Among the innovations in its early years would be the first practical mass-produced four-wheel hydraulic brakes, a system nearly completely engineered by Chrysler with patents assigned to Lockheed, and rubber engine mounts to reduce vibration. Chrysler also developed a road wheel with a ridged rim, designed to keep a deflated tire from flying off the wheel. This safety wheel was eventually adopted by the auto industry worldwide.

Following the introduction of the Chrysler, the Maxwell was dropped after its 1925 model year run, although in truth the new line of lower-priced 4-cylinder Chryslers which were then introduced for the 1926 model year were basically Maxwells which had been re-engineered and rebranded.[24] It was during this time period of the early 1920s that Walter Chrysler assumed the presidency of Maxwell, with the company then ultimately incorporated under the Chrysler name.

Organization

- **Mopar** — Replacement parts for Chrysler-built vehicles. Also comprises **Mopar Performance**, a subdivision providing performance aftermarket parts for Chrysler-built vehicles.
- **Chrysler Financial** — Financial services for Chrysler customers and dealers

Vehicle brands

- **Chrysler** — Passenger cars, minivans and crossovers
- → **Dodge** — Passenger cars, minivans and crossovers
- **Jeep** — SUVs and crossovers
- **Global Electric Motorcars** (GEMCAR) — Battery electric low-speed vehicles
- **RAM** — Trucks and commercial vehicles

Chrysler Headquarters at Auburn Hills

Total American sales

Calendar Year	U.S. Sales	%Chg/yr.
1999[25]	2,638,561	
2000	2,522,695	4.4%
2001[26]	2,273,208	9.9%
2002[27]	2,205,446	3.0%
2003	2,127,451	3.5%
2004[28]	2,206,024	3.7%
2005[28]	2,304,833	4.5%
2006[29]	2,142,505	7.0%
2007[29]	2,076,650	3.1%
2008[30]	1,453,122	30.0%
2009 Jan-Oct[31]	781,319	39%

Vehicles

Turbine

For many years, Chrysler developed gas turbine engines, which are capable of operating on a wide array of combustible fuels, for automotive use. Turbines were common in military vehicles, and Chrysler built many prototypes for passenger cars. In the 1960s, mass production seemed almost ready. Fifty Chrysler Turbine Cars, specialty designed Ghia-bodied coupes were built in 1962 and placed in the hands of consumers for final testing. After further development and testing to make emissions conform to 1970s-enacted EPA standards, the engines were planned as an option for the 1977 model LeBaron. However, Chrysler was forced to abandon the turbine engine as a precondition of U.S. government loan guarantees when the company experienced financial difficulties in the late 1970's

Electric vehicles

Chrysler intends to pursue new drive concepts through ENVI, an in-house organization formed to focus on electric-drive vehicles and related technologies. Established in September, 2007, Chrysler's ENVI division led by Lou Rhodes specifically deals with new all-electric and hybrid vehicles not based on existing models.

Chrysler LLC brought a wide range of green vehicles to the Detroit Auto Show, including three concept vehicles that incorporate electric drive technologies:

1. The Dodge ZEO concept—short for "Zero Emissions Operation"—is an all-electric sport wagon combining a 64-kilowatt-hour lithium-ion battery pack with a 200-kilowatt (268 horsepower) electric motor. The rear-wheel-drive vehicle accelerates to 60 mph (97 km/h) in less than six seconds and has a range of at least 250 miles (400 km). There is also a plug-in hybrid electric version.
2. The Chrysler ecoVoyager concept combines a similar battery pack and motor with a small hydrogen fuel cell to achieve a 300-mile (480 km) range. The vehicle can travel about 40 miles (64 km) on battery power alone and can accelerate to 60 mph (97 km/h) in less than eight seconds.
3. The Jeep Renegade concept, a plug-in hybrid, combines a lithium-ion battery pack with dual 200 kW (270 hp) electric motors on each axle. The Jeep can travel 40 miles (64 km) on battery power alone and can travel 400 miles (640 km) with the help of its 1.5-liter, 3-cylinder clean diesel engine. The vehicle features a lightweight aluminum architecture.

Chrysler is also currently planning at least three hybrid vehicles, the Chrysler Aspen hybrid, Dodge Durango hybrid, and the Dodge Ram hybrid including HEMI engines. Chrysler plans to use hybrid technology developed jointly with General Motors and BMW AG in vehicles beyond the two hybrid SUVs it had already announced to introduce in 2008.[32]

An all-new Ram 1500 pickup will be available as a hybrid in 2010. The Ram HEMI Hybrid will combine a two-mode hybrid system with a 5.7-liter HEMI V-8 engine. For the 2009 Ram 1500, Chrysler is launching an improved version of its HEMI V-8 engine featuring variable valve timing and a four-cylinder mode with an expanded operating range. The result is more power and torque, along with a 4% increase in fuel economy.

Chrysler has also been experimenting with a Hybrid Diesel truck for military applications.

Chrysler has debuted:[33] [34]

- the Dodge EV, an all electric sports car based on the Lotus Europa, with plans for a 120 mph (190 km/h) top speed and a range of 150 to 200 miles (240–320 km).
- plug-in hybrid vehicles (PHEVs), jolting the PHEV mass-production race:[35]
  - the Chrysler EV, a series plug-in hybrid with 40-mile (64 km) all-electric range, based on Chrysler Town & Country.
  - and the Jeep EV, based on a Jeep Wrangler. Chrysler is exploring in-wheel electric motors for this vehicle.

At the 2009 North American International Auto Show in Detroit, Chrysler unveiled the 200C EV Concept, a sports sedan with an all-electric range of 40 miles and an extended range of about 400 miles (640 km). It also added the Jeep Patriot EV, another range-extended electric vehicle. If Chrysler does release an all-electric sports car in 2010, it will be in direct competition with two North American startup companies: Tesla Motors and Fisker Automotive.[36]

Chrysler's ENVI division, which is dedicated to creating production electric drive vehicles, announced in September 2008 that Chrysler LLC will have electric vehicles in showrooms by 2010. They showed three "production intent" vehicles and stated that these are going to be the first of a broad portfolio of electric vehicles.[37]

Chrysler Chief Executive Bob Nardelli said government loans would help speed the electric technology to market. But if they aren't approved, Chrysler will have to spend limited resources on developing new technology and would have to make cuts elsewhere, possibly in employment and development of conventional products. "Unfortunately we have had to furlough many families as a result of the economy turmoil and certainly the downward spiraling in the industry," he said. "I'd like to make sure that we don't have to go further to be able to support advanced technology work."[35]

The Chrysler executives said the day is coming when the whole Chrysler fleet has electric powertrains. "The goal is to achieve fundamental technology, get economies of scale, improve our ability to make the future generations more robust, less cost, smaller, more powerful, better performance," Press said. "Ultimately it will lead to a transformation of our entire fleet that will be in some manner electric drive."[35]

> We chose a technology -- one in which we had the most experience, and which is most accessible to the consumer, and that's electricity.
>
> —*Chrysler CEO Robert Nardelli* [38]

PHEV Research Center

Chrysler is in the Advisory Council of the PHEV Research Center.

Marketing

In 2007, Chrysler began to offer vehicle lifetime powertrain warranty for the first registered owner or retail lessee.[39] The deal covered owner or lessee in U.S., Puerto Rico and The Virgin Islands, for 2009 model year vehicles, and 2006, 2007 and 2008 model year vehicles purchased on or after July 26, 2007. Covered vehicles excluded SRT models, Diesel vehicles, Sprinter models, Ram Chassis Cab, Hybrid System components (including transmission), and certain fleet vehicles. The warranty is non-transferable.[40] However, after Chrysler's restructuring, the warranty program was replaced by five-year/100,000 mile transferrable warranty for 2010 or later vehicles.[41] As of October 5, 2009, Dodge's car and truck line are now split into two, "Dodge" for cars and crossovers and "Ram" for pickup trucks and minivans.[42]

Controversies

Chrysler was among the companies boycotted by gay rights groups after removing advertisements from the ABC sitcom *Ellen* in 1997, which it deemed "controversial."[43]

In 1987, it was discovered that Chrysler sold an estimated 32,750 cars that had been test-driven with disconnected odometers - some as much as 500 miles - before being shipped to dealers. Chrysler settled out of court with complainants.[44] [45] Chrysler CEO Lee Iacocca sought to minimize damage to the corporation's public image by calling a news conference in which he termed the action "dumb" and "unforgivable".

See also

- List of automobile manufacturers
- American Motors Corporation
- Chrysler Building
- Hemi engine
- Lee Iacocca
- Mopar
- Chrysler Headquarters and Technology Center
- Chrysler Proving Grounds
- List of Chrysler factories

Chrysler 51

Countries

- Chrysler Australia
- Chrysler Fevre Argentina

External links

- Chrysler LLC corporate website [46]
- Chrysler Official UK brand site [47]
- "Granholm Wants Federal Funds for Carmakers" [48]

References

[1] Chrysler LLC (2009-05-20). " C. Robert Kidder to Become Chairman of Chrysler Group LLC (http://www.examiner.com/p-349852-C__Robert_Kidder_to_Become_Chairman_of_Chrysler_Group_LLC.html)". Press release. Archived from on 2009-05-20. . Retrieved 2009-06-09.
[2] " UPDATE: US Ruling Paves Way For Marchionne As Chrylser [sic (http://online.wsj.com/article/BT-CO-20090601-708155.html) CEO]". Dow Jones. Dow Jones & Company, Inc.. 2009-06-01. . Retrieved 2009-06-09.
[3] " Fiat Said to Buy Chrysler Assets Today to Form New Automaker (http://www.bloomberg.com/apps/news?pid=newsarchive&sid=aAB9jCmPBUQU)". www.bloomberg.com. . Retrieved 2009-06-11.
[4] Chrysler Group LLC list of company brands (http://www.chryslergroupllc.com/company/brands) retrieved on 18-November 2009.
[5] http://www.chryslergroupllc.com/
[6] " Chrysler Reviews and History (http://www.jbcarpages.com/chrysler/)". JB car pages. . Retrieved 2008-09-22.
[7] " Chrysler History (http://www.jbcarpages.com/chrysler/history/)". JB car pages. . Retrieved 2008-09-22.
[8] " Cerberus Takes Majority Interest in Chrysler Group and Related Financial Services Business for EUR 5.5 Billion ($7.4 billion) (http://www.daimlerchrysler.com/dccom/0-5-7145-1-858191-1-0-0-0-0-0-11979-0-0-0-0-0-0-0-0.html)". DaimlerChrysler. .
[9] " Cerberus gains control of Chrysler (http://www.mercurynews.com/arts/ci_6534916)". San Jose Mercury News. .
[10] " Daimler Reaches Agreement On Separation From Chrysler (http://online.wsj.com/article/BT-CO-20090427-717397.html)". 27 April 2009. .
[11] Chrysler bankruptcy plan is announced (http://www.nytimes.com/2009/05/01/business/01auto.html?_r=1&hp). www.nytimes.com. Retrieved 2009-04-30.
[12] " Court Approves Sale of Chrysler LLC Operations to New Company Formed with Fiat (http://news.prnewswire.com/DisplayReleaseContent.aspx?ACCT=104&STORY=/www/story/06-01-2009/0005035600&EDATE=)". News.prnewswire.com. . Retrieved 2009-06-06.
[13] " Chrysler Gets OK for Fiat Sale (http://www.edmunds.com/insideline/do/News/articleId=149327)". Edmunds.com. 2009-06-01. . Retrieved 2009-06-06.
[14] Ramsey, Mike and Kary, Tiffany. "Chrysler Assets Said to Have Little Net Proceeds for Creditors" Bloomberg.com, 2009-06-23 (http://www.bloomberg.com/apps/news?pid=20601103&sid=aBhumEPf5wbM), retrieved on 2009-07-10.
[15] de la Mercel, Michael; Micheline Maynard (June 10, 2009). " Swift Overhaul Moves Ahead as Fiat Acquires Chrysler Assets (http://www.nytimes.com/2009/06/11/business/global/11chrysler.html)". New York Times. . Retrieved June 10, 2009.
[16] Forden, Sara Gay; Mike Ramsey (June 10, 2009). " Fiat Said to Buy Chrysler Assets Today to Form New Automaker (http://www.bloomberg.com/apps/news?pid=20601103&sid=a59MFT3OCDBs)". Bloomberg. . Retrieved June 10, 2009.
[17] Davis, Mike; Tell, David (1995). *The Technology Century: 100 years of The Engineering Society 1895-1995*. Engineering Society of Detroit. p. 53. ISBN 9781563780226.
[18] " A Brief Look at Walter P. Chrysler (http://www.chryslerclub.org/walterp.html)". WPC News. .
[19] Malis, Carol (1999). *Michigan: celebrating a century of success*. Cherbo Publishing Group. p. 76. ISBN 9781882933235.
[20] Kimes, Beverly Rae; and Clark, Jr., Henry Austin (1989). *Standard Catalog of American Cars 1805-1942 (2nd ed.)*. Krause Publications. p. 292. ISBN 0873411110.
[21] Kimes, Beverly Rae; and Clark, Jr., Henry Austin (1989). *Standard Catalog of American Cars 1805-1942 (2nd ed.)*. Krause Publications. p. 257. ISBN 0873411110.
[22] Kimes, Beverly Rae; and Clark, Jr., Henry Austin (1989). *Standard Catalog of American Cars 1805-1942 (2nd ed.)*. Krause Publications. pp. 1498. ISBN 0873411110.
[23] Zatz, David. " Chrysler Technological Innovations (http://www.allpar.com/corporate/technology.html)". . Retrieved 2008-01-05.
[24] Kimes, Beverly Rae; and Clark, Jr., Henry Austin (1989). *Standard Catalog of American Cars 1805-1942 (2nd ed.)*. Krause Publications. pp. 292-293, 901. ISBN 0873411110.
[25] " Chrysler Group Announces Year-End and December Sales (http://www.theautochannel.com/news/press/date/20010103/press033497.html)". Theautochannel.com. . Retrieved 2009-04-30.

[26] " Chrysler Group Reports U.S. December Sales (http://www.theautochannel.com/news/2003/01/04/152240.html)". Theautochannel.com. . Retrieved 2009-04-30.
[27] " Chrysler Group Reports December 2003 Sales Increase of 2 Percent (http://www.theautochannel.com/news/2004/01/05/175827.html)". Theautochannel.com. 2004-11-17. . Retrieved 2009-04-30.
[28] " Chrysler Group 2005 U.S. Sales Rise 5 Percent, Highest Since 2000; December Sales Decline In Line with Overall Industry (http://www.prnewswire.com/cgi-bin/stories.pl?ACCT=104&STORY=/www/story/01-04-2006/0004242655&EDATE)". Prnewswire.com. . Retrieved 2009-04-30.
[29] " Total Chrysler LLC December 2007 Sales Up 1 Percent on the Strength of Retail; Demand... (http://www.reuters.com/article/pressRelease/idUS217066+03-Jan-2008+PRN20080103)". Reuters. 2008-01-03. . Retrieved 2009-04-30.
[30] " Chrysler LLC Reports December 2008 U.S. Sales (http://news.prnewswire.com/DisplayReleaseContent.aspx?ACCT=104&STORY=/www/story/01-05-2009/0004949198&EDATE=)". News.prnewswire.com. . Retrieved 2009-04-30.
[31] " Chrysler Group LLC Reports October 2009 U.S. Sales Increase Compared with (http://www.prnewswire.com/news-releases/chrysler-group-llc-reports-october-2009-us-sales-increase-compared-with-september-2009-68887307.html)". Michigan: Prnewswire.com. . Retrieved 2009-11-23.
[32] Planet Ark : Chrysler Commits to New Hybrids, Better Mileage (http://www.planetark.com/dailynewsstory.cfm/newsid/42750/story.htm)
[33] Neff, John (2008-09-23). " Chrysler LLC debuts Dodge EV, Jeep EV and Chrysler EV (http://www.autoblog.com/2008/09/23/chrysler-llc-debuts-dodge-ev-jeep-ev-and-chrysler-ev/)". Autoblog.com. . Retrieved 2009-04-30.
[34] " EERE News: EERE Network News (http://apps1.eere.energy.gov/news/enn.cfm#id_11990)". Apps1.eere.energy.gov. . Retrieved 2009-04-30.
[35] " Chrysler "Jolts" PHEV Race; PHEV Ads; V2Green Acquired (http://www.calcars.org/calcars-news/1005.html)". Calcars.org. . Retrieved 2009-04-30.
[36] " EERE News: Chrysler, Ford, and Other Automakers Pursue Electric Vehicles (http://apps1.eere.energy.gov/news/news_detail.cfm/news_id=12178)". Apps1.eere.energy.gov. 2009-01-14. . Retrieved 2009-04-30.
[37] " Innovation − ENVI (http://www.chryslerllc.com/en/innovation/envi/overview/)". Chrysler LLC. 2008-09-22. . Retrieved 2009-04-30.
[38] " Developments on Bridge Loan to Automakers; 4 Groups Write Congress (http://www.calcars.org/calcars-news/1028.html)". Calcars.org. . Retrieved 2009-04-30.
[39] BREAKING: Chrysler announces lifetime powertrain warranty! (http://www.autoblog.com/2007/07/26/breaking-chrysler-announces-lifetime-powertrain-warranty/)
[40] NEW CHRYSLER LIFETIME POWERTRAIN WARRANTY CUSTOMERS -- Q&A (http://www.chrysler.com/en/lifetime_powertrain_warranty/faq.html)
[41] REPORT: Chrysler dropping lifetime powertrain warranty to five-year/100,000 miles (http://www.autoblog.com/2009/08/20/report-chrysler-dropping-lifetime-powertrain-warranty-to-5-year/)
[42] (http://www.google.com/hostednews/ap/article/ALeqM5j_RSDTBP19jDsE1Zo1XR84LhLDbgD9B4ULKG1)
[43] Gallagher, John (1997-06-10). " The ad buck stops here - controversy on gays and advertising (http://findarticles.com/p/articles/mi_m1589/is_n735/ai_20164880)". The Advocate. . Retrieved 2007-11-07.
[44] Chrysler Fined for Safety Violations (http://query.nytimes.com/gst/fullpage.html?res=9B0DEEDF133EF931A25754C0A961948260)*New York Times*, 12 Jul 1987
[45] Chrysler to Pay Some 40,000 Owners in Settlement (http://query.nytimes.com/gst/fullpage.html?res=940DE2DC113DF937A1575BC0A96E948260)*New York Times*, Aug 24 1988
[46] http://www.chryslergroupllc.com
[47] http://www.chrysler.co.uk/
[48] http://www.wxyz.com/news/story.aspx?content_id=547e9947-be50-4ace-9a05-9692c0a50d2b

Muscle car

Muscle car is a term used to refer to a variety of high performance → automobiles.[1] [2] At its most widely accepted the term refers to American 2-door rear wheel drive mid-size cars of the late 1960s and early 1970s equipped with large, powerful V8s and sold at an affordable price for street use and drag racing, formally and informally.[3] [4] [5] [6]

This 1966 Pontiac GTO is an example of classic muscle.

As such, they are distinct from two-seat sports cars and expensive 2+2 GTs intended for high-speed touring and road racing.

Building on the American phenomenon and developing simultaneously in their own markets, muscle cars also emerged in their own fashions in Australia, South Africa, the UK and elsewhere.

Definition

Hemi-powered 1970 Plymouth Road Runner.

Though the notion of a *Muscle car* as an American two-door with a big engine sold at an affordable price for street and drag racing is generally held, there is much blurring around its edges.

According to a contemporary issue of *Road Test* magazine (June 1967), a "muscle car" is "Exactly what the name implies. It is a product of the American car industry adhering to the hot rodder's philosophy of taking a small car and putting a BIG engine in it [...] The Muscle Car is Charles Atlas kicking sand in the face of the 98 hp (73 kW) weakling."[7] Author of the book *Muscle Cars* the quote is drawn from, Peter Henshaw, furthers that the muscle car was designed for straight-line speed, and did not have the "sophisticated chassis", "engineering integrity" or "lithe appearance" of European high-performance cars[7]

Opinions vary as to whether high-performance full-size cars, compacts, and pony cars qualify as muscle cars.[8]

The following is a list of muscle cars and their manufacturers (along with the pony car of the same company):

Manufacturer	Pony car	→ Muscle car
AMC	Javelin	AMX
Chevrolet	Camaro	Chevelle
→ Dodge	Challenger	→ Charger
Ford	Mustang	Torino
Mercury	Cougar	Montego
Oldsmobile	*none*	442
Plymouth	Barracuda	Road Runner
Pontiac	Firebird	GTO

Development

Early muscle

Opinions on the origin of the muscle car vary, but the 1949 Oldsmobile Rocket 88, created in response to public interest in speed and power, is often cited as the first of the breed. It featured an innovative and powerful new engine—America's first high- compression overhead valve V-8—in the lighter Oldsmobile body.[9]

Musclecars magazine wrote: "[t]he idea of putting a full-size V8 under the hood of an intermediate body and making it run like Jesse Owens in Berlin belongs to none other than Oldsmobile... [The] all-new ohv V8...Rocket engine quickly found its way into the lighter 76 series body, and in February 1949, the new 88 series was born."[10]

1949 Rocket 88 engine

The article continued: "Walt Woron of Motor Trend enjoyed the 'quick-flowing power...that pins you to your seat and keeps you there until you release your foot from the throttle [...] Olds dominated the performance landscape in 1950, including wins in the NASCAR Grand National division, Daytona Speed Weeks, and the 2100-plus-mile Carrera Panamericana. In France, an 88 won a production car race at Spa-Francorchamps... A husky V8 in a cleanly styled, lightweight coupe body, the original musclecar truly was the '49 Olds 88."[10]

Hudson Hornet: Rocket 88's only competitor

Jack Nerad wrote in *Driving Today*: "the Rocket V-8 set the standard for every American V-8 engine that would follow it for at least three decades [...] With a displacement of 303 cubic inches and topped by a two-barrel carburetor, the first Rocket V-8 churned out 135 horsepower (101 kW) at 3,600 rpm and 263 pound-feet of torque at a lazy 1800 rpm [and] no mid-range car in the world, save the Hudson Hornet, came close to the Rocket Olds performance potential..."

Nerad added that the Rocket 88 was "the hit of NASCAR's 1950 season, winning eight of the 10 races. Given its lightning-like success, one could clearly make the case that the Olds 88 with its 135 horsepower (101 kW) V-8 was the first 'musclecar'..."[11]

Steve Dulcich, writing in *Popular Hot Rodding*, also cites Oldsmobile, concurrently with Cadillac, as having "launched the modern era of the high-performance V-8 with the introduction of the "Rocket 88" overhead-valve V-8 in 1949."[12]

Growth of the trend

Other manufacturers "showcased performance hardware in flashy limited-edition models. Chrysler led the way with its 1955 C-300, an inspired blend of Hemi power and luxury-car trappings that fast became the new star of NASCAR. With 300 horsepower (224 kW), it was rightly advertised as 'America's Most Powerful Car.'"[9]

Capable of accelerating from 0 to 60 mph (97 km/h) in 9.8 seconds and reaching 130 miles per hour (209 km/h), the 1955 Chrysler 300 is also recognized as one of the best-handling cars of its era.[13]

1955 Chrysler C-300, "America's most powerful car", had 300 horsepower (220 kW).

America's fastest 1957 sedan: Rambler Rebel had lightweight unibody construction and V8 engine.

Two years later the Rambler Rebel was the fastest stock American sedan, according to *Motor Trend*.[14] The popularity and performance of muscle cars grew in the early 1960s, while Mopar (Dodge, Plymouth, and Chrysler) and Ford battled for supremacy in drag racing—the 1962 Dodge Dart 413 cu in (6.8 L) Max Wedge, for example, could run a 13-second 1/4-mile dragstrip at over 100 miles per hour (161 km/h). By 1964, there were Oldsmobile, Chevrolet, and Pontiac muscle cars in GM's lineup, and Buick joined them a year later. For 1964 and 1965, Ford had its 427 cu in (7 L) Thunderbolts, and Mopar unveiled the 426 cu in (7 L) Hemi engine. The Pontiac GTO was an option package that included Pontiac's 389 cu in (6.4 L) V8 engine, floor-shifted transmission with Hurst shift linkage, and special trim. In 1966 the GTO became a model in its own right. The project, spearheaded by Pontiac division president John DeLorean, technically violated GM's policy limiting its smaller cars to 330 cu in (5.4 L) displacement, but the new model proved more popular than expected and inspired GM and its competitors to produce numerous imitators. The GTO itself was a response to the Dodge Polara 500 and the Plymouth Sport Fury, which in 1962 had been shrunk to intermediates—at a time when bigger was considered better.

AMC, though late entering the muscle car market, produced "an impressive array of performance cars in a relatively short time," said *Motor Trend*. "The first stirrings of AMC performance came in 1965, when the dramatic if ungainly Rambler Marlin fastback was introduced to battle the Ford Mustang and Plymouth Barracuda."[15] Although the Marlin was a flop in terms of sales and initial performance, AMC gained some muscle-car credibility in 1967, when it made both the Marlin and the "more pedestrian" Rebel available with its new 280 horsepower (209 kW), 343 cu in (5.6 L) "Typhoon" V8. And in 1968 the company offered two legitimate muscle car contenders: the Javelin and its truncated variant, the AMX.[15]

Although the sales of true muscle cars were relatively modest by total Detroit production standards, they had value in publicity and bragging rights. Competition between manufacturers meant that buyers had the choice of ever-more powerful engines—a horsepower war that peaked in 1970, with some models offering as much as 450 hp (336 kW) (with this and others likely producing as much or more actual power, whatever their rating).

Turn-key drag racers

Muscle cars attracted young customers (and their parents) into showrooms, and they bought the standard editions of these mid-size cars. To enhance the "halo" effect of these models, the manufacturers modified some of them into turn-key drag racers.

For example, Ford built 200 lightweight Ford Galaxies for drag racing in 1963. All non-essential equipment was omitted. Modifications included fiberglass panels, aluminum bumpers, traction bars, and a competition-specification 427 cu in (7 L) engine factory-rated at a conservative 425 bhp (317 kW). This full-size car could run the quarter mile in a little over 12 seconds.[16] Also built in 1963 were 5,000 road-legal versions that could be used every day. (Ford claimed 0-60 in less than 6 seconds for the similarly-powered 1966 Galaxie 500XL 427.)[7]

Another Ford lightweight was the 1964 Ford Thunderbolt that utilized the mid-size Fairlane body. A stock Thunderbolt could run a quarter-mile (402 m) at a drag strip in 11.76 seconds at 122.7 mph (197.5 km/h),[17] and Gas Ronda dominated the NHRA World Championship with a best time of 11.6 seconds at 124 mph (200 km/h).[7] The Thunderbolt included competition-specification 427 cu in (7 L) engine and special exhausts (though technically legal for street use, the car was too raucous for the public roads—"not suitable", according to a *Hot Rod* magazine quote, "for driving to and from the strip, let alone on the street in everyday use";[17] also massive traction bars, asymmetrical rear springs, and a trunk-mounted 95-pound (43 kg) bus battery to maximize traction from what was realistically 500 bhp (373 kW).[17] Sun visors, exterior mirror, sound-deadener, armrests, jack, and lug wrench were omitted to save weight. The car was given lightweight Plexiglass windows, and early versions had fiberglass front body panels and bumpers, later changed to aluminum to meet NHRA regulations.[18] Base price was US$3,780.[17] 111 Thunderbolts were built, and Ford contracted Dearborn Steel Tubing to help with assembly. Factory records show that the first 11 cars were maroon and the subsequent 100 were white.[19]

Road-legal drag racer: with 427 V8 in lightened midsize Ford Fairlane body, stock 1964 Ford Thunderbolt ran 11.76-second quarter mile.

The 1964 Dodge 426 Hemi Lightweight produced over 500 bhp (370 kW). This "top drag racer" had an aluminium hood, lightweight front bumpers, fenders, doors and lower valance, magnesium front wheels, lightweight Dodge van seat, Lexan side windows, one windshield wiper and no sun visors or sound deadening. Like other lightweights of the era it came with a factory disclaimer: *Designed for supervised acceleration trials. Not recommended for general everyday driving because of the compromises in the all-round characteristics which must be made for this type of vehicle.*[18]

Also too "high-strung" for the street was Chrysler's small-volume-production 1965 drag racer, the 550 bhp (410 kW) Plymouth Satellite 426 Hemi. Although the detuned 1966 version (the factory rating underestimated it at 425 bhp (317 kW)) has been criticized for poor brakes and cornering, *Car and Driver* described it as "the best combination of brute performance and tractable street manners we've ever driven." The car's understated appearance belied its "ultra-supercar" performance: it could run a 13.8-second quarter mile at 104 mph (167 km/h). Base price was $3,850.[20]

Chevrolet likewise eschewed flamboyant stripes and badges for their 1969 Chevelle COPO 427 and kept its appearance low-key. The car could run a 13.3 sec. quarter-mile at 108 mph (174 km/h). Chevrolet rated the engine at 425 hp (317 kW), but the NHRA claimed a truer 450 hp (340 kW).[21] It has been said that the 1969 COPO Chevelles were "among the most feared muscle cars of any day. And they didn't need any badges."[21] Base price was US$3,800.[21]

For 1970 Chevrolet offered the Chevelle SS 454, also at a base price of US$3,800. The "muscle car summit", its 454 cu in (7.4 L) engine was rated at 450 hp (336 kW), the highest-ever factory rating at that time. *Car Life* magazine wrote: "It's fair to say that the Supercar as we know it may have gone as far as it's going."[22]

Youth market and "budget muscle"

The general trend towards higher performance in factory-stock cars reflected the importance of the youth market. A key appeal of muscle cars was that they offered the burgeoning American car culture relatively affordable and powerful street performance in models that could also be used for drag racing. But as size, optional equipment and luxury appointments increased, engines had to be more powerful to maintain performance levels, and the cars became more expensive.

1970 Plymouth GTX 440: base price US$3,355 and "more performance per dollar" than most other cars of its time.

In response to rising cost and weight, a secondary trend towards more basic "budget" muscle cars emerged in 1967 and 1968—e.g. the "original budget Supercar"[23] Plymouth Road Runner; also the Plymouth GTX, which offered "as much performance-per-dollar as anything on the market, and more than most",[24] the Dodge Super Bee and other variants. Manufacturers also offered bigger engines in their compact models, sometimes making them lighter, roomier, and faster than their own pony-car lines.

The 340 cu in (5.6 L)-powered 1970 Plymouth Duster was one of these smaller, more affordable cars. Based on the compact-sized Plymouth Valiant and priced at US$2,547, the 340 Duster posted a 6.0-second 0-60 mph (97 km/h) time and ran the quarter mile in 14.7 seconds at 94.3 mph (151.8 km/h).[25] This "reasonably fast" compact muscle car had a stiff, slightly lowered suspension which, in the view of *Hot Rod* magazine at the time, let the car "ride in an acceptable fashion".[26] However an anonymous 2007 article on the *Consumer Guide* website refers to "a punishing ride" and trim that was "obviously low-budget."[25] The 1970 model came with front disc brakes and without hood scoops. The only high-performance cues were dual exhausts and modest decals.[25] Tom Gale, former Chrysler vice president of design, describes the car as "a phenomenal success. It had a bulletproof chassis, was relatively lightweight, and had a good power train. These were 200000-mile (320000 km) cars."[27] *Hot Rod* rated the Duster "one of the best, if not the best, dollar buy in a performance car" in 1970.[26]

American Motors' mid-sized 1970 Rebel Machine, developed in consultation with Hurst Performance, was also built for normal street use. It had a 390 cu in (6.4 L) engine developing 340 hp (254 kW) — a "moderate performer"[28] that gave a 0-60 mph (97 km/h) time of 6.8 seconds and a quarter mile in 14.4 seconds at 99 mph (159 km/h).[29] Early examples came in "patriotic" red, white and blue.[30] Jack Nerad wrote in Driving Today that it was "a straight-up competitor to the GTO, et al. ... [T]he engine was upgraded to 340 horsepower (250 kW) [with] a four-barrel Motorcraft carburetor and other hot rod trickery. The torque figure was equally prodigious—430 pound-feet at a lazy 3600 rpm. In this car the engine was practically the entire story." With four-speed manual transmission, the car "could spring from zero to 60 miles per hour in just 6.4 seconds..." In Nerad's view the car "somehow, someway deserves to be considered among the Greatest Cars of All Time."[31]

"The Machine": factory-modified 1970 AMC Rebel ran 14.4-second quarter mile in stock trim.

A post-2005 *Mopar Muscle* magazine article said, "But by far the most stunning thing for a car with this level of performance and standard equipment was the sticker of just US$3,475."[32] In 1970, *Hot Rod* magazine wrote: "Here's a car that lists for $3500 at the starting point, but lacks an appealing interior, feels way too big (and is) to be a handler, and is marked with more identity than Peter Fonda's two wheeler,[33] with about the same taste. Not many of the folks we talked with while we had the car could think of any reason they'd want this car, with 36 months to pay and all the bright paint." The author said, "[I]f there is an attempt here to chase down the well-known middle-class supercar market nobody but American Motors need worry."[34]

For comparison, the "plain wrapper"[35] 1969 Plymouth Road Runner, *Motor Trend* magazine's Car of the Year, ran a 14.7 quarter at 100.6 mph (161.9 km/h) with the standard 383 cu in (6.3 L) engine after the addition of a high-performance factory camshaft plus non-standard, high-performance induction and exhaust manifolds, carburetor and slick tires. In this form the car cost US$3,893.[23] In 1968 Dodge's US$3,027 Super Bee ran a 15-second quarter at 100 mph (160 km/h) on street tires with the same engine, only stock.[36]

Furthermore, the 340 cu in (5.6 L)-powered 1968 Plymouth Barracuda 4-seater, which *Hot Rod* magazine categorized as "a supercar, without any doubt attached...also a 'pony car', a compact and a workhorse" with enough rear seat leg and head room for "passengers to ride back there without distress" and "a flip-up door to the trunk area for ferrying some pretty sizeable loads of cargo", was a "sizeable threat on the drag strip": 13.33 seconds at 106.50 mph (171.40 km/h). Base price was $2796.00. Price as tested by *Hot Rod*: $3652.[37]

Related pickup trucks

Another related type of vehicle is the car-based pickup (known colloquially in Australia as a "ute" (short for "utility"). Holden and Ford Australia make such a vehicle under the model name "Ute""). Examples of these are the Ford Ranchero, GMC Sprint, GMC Caballero, and one of the most famous examples, the Chevrolet El Camino.

Decline

The automotive safety lobby led by Ralph Nader decried offering powerful cars for public sale, particularly when targeted at young buyers: the power of many muscle cars underlined their marginal brakes, handling, and tire adhesion. In response, the automobile insurance industry levied surcharges on all high-powered models, an added cost that put many muscle cars out of reach of their intended buyers. Simultaneously, efforts to combat air pollution—a problem that grew more complicated in 1973 when the OPEC oil embargo led to price controls and gasoline rationing—focused Detroit's attention on emissions control.

A majority of musclecars came optioned with high-compression powerplants - some as high as 11:1. Prior to the oil embargo, 100-octane fuel was common (e.g. Sunoco 260, Esso Extra, Chevron Custom Supreme, Super Shell, Texaco Sky Chief, Amoco Super Premium, Gulf No-nox) until the passage of the Clean Air Act of 1970 where octane ratings were lowered to 91 - due in part of the removal of tetraethyl lead as a valve lubricant. Unleaded gasoline was phased in.

With all these forces against it, the market for muscle cars rapidly evaporated. Horsepower began to drop in 1971 as engine compression ratios were reduced. High-performance engines like Chrysler's 426 Hemi were discontinued, and all but a handful of other performance models were discontinued or transformed into soft personal luxury cars. Some nameplates e.g. Chevrolet's SS or Oldsmobile's 442 would become sport appearance packages (known in the mid to late 1970s as the vinyl and decal option - Plymouth's Road Runner was an upscale decor package for their Volare coupes). One of the last to succumb, a car that *Car and Driver* dubbed "The Last of the Fast Ones", was Pontiac's Trans Am SD455 model of 1973–1974. In 1975 its performance was markedly reduced, although it remained in production through 2002 and was made powerful again from 1993 onwards.

American performance cars began to make a return in the 1980s. Owing to increases in production costs and tighter regulations governing pollution and safety, these vehicles were not designed to the formula of the traditional low-cost muscle cars. The introduction of electronic fuel injection and overdrive transmission for the remaining 1960s muscle-car survivors—the Ford Mustang, Chevrolet Camaro and Pontiac Firebird—helped sustain a market share for them alongside personal luxury coupes with performance packages, i.e. the Buick Regal T-Type or Grand National, Ford Thunderbird Turbo Coupe and Chevrolet Monte Carlo SS circa 1983-88.

Australia

Australian muscle: 1970 Holden HG Monaro GTS 350 V8

Australia developed its own muscle car tradition around the same period, with the big three manufacturers Ford Australia, Holden or Holden Dealer Team (by then part of General Motors), and Chrysler Australia. The cars were specifically developed to run in the Armstrong 500 (miles) race and later the Hardie Ferodo 500 (the race's current 1,000 kilometre format was adopted in 1973). The demise of these cars was brought about by a change in racing rules requiring that 200 examples had to be sold to the general public before the car could qualify (homologation). In 1972, the government stepped in to ban supercars from the streets after two notable cases. The first instance was a *Wheels* magazine journalist driving at 150 mph (240 km/h) in a 1971 Ford Falcon GTHO Phase III XY 351.[38] Whilst the car was getting exposure in the press, the second incident occurred in George Street, Sydney, when a young male was caught driving at an estimated 150 mph (240 km/h) through the busy street, in a 1971 Ford Falcon GTHO Phase III, drag racing a Holden Monaro GTS 350. This was known in Australia as "The Supercar Scare".

Ford produced what is considered to be the first Australian muscle car in 1967, the 287 cu in (4.7 L) Windsor-powered Ford Falcon GT XR. Ford continued to release faster models, culminating in the Ford Falcon GTHO Phase III of 1971, which was powered by a factory modified 351 Cleveland. Along with its GT and GTHO models, Ford, starting with the XW model in 1969, introduced a 'sporty' GS model, available across the Falcon range. The basic GS only came with a 250 cu in (4.1 L) six cylinder engine, but the 302 cu in (4.9 L) and 351 cu in (5.8 L) Windsor (replaced by the Cleveland engines for the XY), were optional. Ford's larger, more luxurious Fairlane was also available with these engines and could also be optioned with the 300 bhp (224 kW) 351 cu in (6 L) "Cleveland" engine.

Holden produced the famous Holden Monaro with 307 cu in (5 L), 327 cu in (5.4 L), and 350 cu in (5.7 L) Chevrolet smallblocks or 253 cu in (4.1 L) and 308 cu in (5 L) Holden V8s, followed by the release of four high-performance Toranas, the GTR-XU1 (1970–1973), SL/R 5000 (1974–1977), L34 (1974) and the A9X (1977).

The XU1 Torana was originally fitted with a 186 cu in (3 L) triple carbureted 6-cylinder engine, later increased to 202 cu in (3.3 L), as opposed to the 308 cu in (5 L) single quad-barrel carbureted V8 in the SL/R 5000, L34, and A9X.

Ford Falcon Cobra 351 V8

Chrysler produced the R/T Valiant Charger from 1971 to 1973 when the R/Ts were discontinued; the dominant R/T models were the E38 and E49 with high performance 265 cu in (4.3 L) Hemi engines featuring triple Weber carburetors.

Chrysler apparently considered a high-performance V8 program importing 338 340 cu in (5.6 L) V8 engines from the U.S.

That high-performance project never went ahead, and the engines were subsequently fitted to the upmarket 770 model Charger. Initially this model was designated "SE" E55 340 (V8) and only available with automatic transmission; with a model change to the VJ in 1973 the engine became an option, and the performance was lessened.

All Chrysler performance Chargers were discontinued in 1974 with the exhausting of high performance 265ci Hemi and 340 V8s.

The Australian muscle car era is considered to have ended with the release of the Australian Design Rule regarding emissions in ADR27a in 1976. An exception to this rule was the small number of factory-built Bathurst 1000 homologation specials that were constructed after 1976: they are considered to be musclecars. Examples of these homologation specials include the Torana A9X and the Bathurst Cobras.

Later homologation cars were built outside of the factory, many by the Holden Dealer Team (HDT) for track and road use. Although not regarded as true muscle cars, they quickly gained an enthusiastic following. The HDT program was under Peter Brock's direction and had approval from Holden.

Several highly modified high-performance road-going Commodores were produced through the early and mid 1980s. These "homologation specials" were produced to meet the Group A racing regulations. Models included the VC Group C, the VH SS Group III with a 0-100 km/h of 6.7 seconds,[39] the Blue VK SS Group A and the burgundy VL SS Group A. These vehicles are all individually numbered with only 4246 Brock HDT's made and are considered to be collectors' items.

The HDT Commodores are highly collectible muscle cars. Holden Dealer Team vehicles' became more collectible than ever in the wake of Brock's 2006 death.

Showroom-condition HDT cars are generating prices as high as $200,000 AU.[40]

South Africa

In South Africa, Chevrolet placed the Z28 302 Chevrolet smallblock into a Vauxhall Viva coupe bodyshell and called it the Firenza CanAm. Basil Green produced the 302 Windsor–powered Capri Perana. In addition Australian HT and HG GTS Monaros (1969-71) were exported in CKD form and were given a new fascia and rebadged as a Chevrolet SS, which were sold until about 1973. Falcon GTs were also exported to South Africa and rebadged as Fairmont GTs. The Australian XW Falcon GT was called the 1970 Fairmont GT, and the XY Falcon GT was called the Fairmont GT. The Falcons were re-badged as Fairmonts because of to the bad reputation of the American Falcons at the time. The Fairmonts were almost the same as their Australian cousins apart from a few cosmetic differences.

United Kingdom

In the United Kingdom, the muscle car never gained a significant market, but it certainly influenced British manufacturers, with models such as the Ford Capri and Vauxhall Firenza directly inspired by American designs. Ford did, however, fit the Ford Capri with a powerful V6 Engine, introducing the muscle car principle of affordable power to the British public. The Dagenham built 2944cc Ford Capri 3000GT (or 3000E in more luxurious trim), together with the Ford Capri 2600GT produced at Ford's plant in Cologne, were both muscle cars in spirit. Although the Ford Capri was, in essence, a pony car, GM responded with the Opel Commodore Coupe which was based on the mid-size Opel Rekord. In GTE form and fitted with a six cylinder engine, it was able to match the Fords for performance. Later, both Ford and Vauxhall continued the tradition of producing high performance variants (known colloquially as Q-cars) of its family cars, though these tended to have subtle styling - the effect being to create a discreet performance car.

450 bhp Cosworth V6-powered Ford Capri ETC (European Touring Car Championship) race car

Modern muscle cars

America and Australia

In the U.S., the full-size, 4-door Chevrolet Impala SS had a short but popular production run from 1994–1996 as a high-performance limited-edition version of the Caprice equipped with a Corvette-derived 5.7 L V8 LT1 engine and other specific performance features and body styling using the options found on the Caprice 9C1 police package. The revived Impala SS was no match for the rising sport utility market; some analysts would consider GM's phasing out rear-wheel drive luxury sedans as a fatal mistake.

1994-1996 Chevrolet Impala SS

The Impala SS nameplate was resurrected again in 2003 as a high-performance version of the standard Impala with larger and/or supercharged engines (whether the 21st century Impalas, which are front-wheel drive, Canadian-built, and have had variously V6s and V8s, can be considered muscle cars in the same vein as their earlier namesakes is debatable). General Motors discontinued its F-body pony-car models, the Chevrolet Camaro and Pontiac Firebird after 2002 but brought back the GTO in 2004 as a rebadged Holden Monaro imported from Australia.

Sales were poor and the "new" GTO was discontinued after three years. In its first sales year it achieved only 14,000 of its 18,000 per-annum sales target. The styling was unpopular—*Car and Driver* described the GTO as "[l]usty performance disguised in a phone-company fleet car"—and the already sluggish sales fell to 11,600 in 2006, its last model year.[41] However, according to an April 2008 article in *Car and Driver*, GM remained "undaunted" in its plan to "pull Pontiac's performance bona fides out of mothballs using the next generation of Australian-engineered-and-built rear-drivers. The agenda includes the G8 sedan..." [42] and also a Camaro, "to be built in Canada at the plant that builds the Buick LaCrosse and Pontiac Grand Prix."[43]

2003-2004 Mercury Marauder

For 2003 and 2004 Mercury revived its old Marauder nameplate, as a modified 302 hp (225 kW) Mercury Grand Marquis (based on the Ford Crown Victoria Police Interceptor). The Marauder failed to attract a market share unlike GM's Impala SS revival. In 2005 a "retro-inspired" version of the pony car Ford Mustang went on sale, which drew various design cues from Mustangs of the mid to late 1960s and early-1970s. In 2007 Ford and Shelby also re-released a new and modern version of the G.T. 500, with Super Snake and King of the Road editions following closely behind in 2008. Saleen has introduced a special edition based on the classic BOSS Mustangs of 1970 called the "S302 Parnelli Jones" after a famous Trans-Am series driver from the 1960s and 1970s, Parnelli Jones (a subsequent similar model followed with Dan Gurney's namesake).

In 2004 Chrysler introduced their LX platform, which serves as the base for a new line of rear-wheel drive, V8-powered cars (using the new Hemi engine), including a four-door version of the Dodge Charger. While purists would not consider a station wagon (the Dodge Magnum) or a four-door sedan a muscle car, the performance of the new models is the equal of many of the vintage muscle cars of legend. Dodge has also revived two "classic" model names with the Charger: Daytona and Super Bee. The first was featured in 2006 as a → Dodge Charger Daytona R/T and the Super Bee joined in 2007 as the Dodge Charger Super Bee. In addition, Dodge has been developing a new performance vehicle under the Challenger badge, which borrows styling cues from its older namesake, the prototype for which made its debut at the 2006 North American International Auto Show. Chevrolet has recently unveiled their Camaro concept car as well, with plans to sell new Camaros beginning with the 2009 model year.

GM's Cadillac division, which has marketed luxury cars for decades, introduced its XLR roadster in 2004 currently produced alongside the Chevrolet Corvette in its Bowling Green, Kentucky manufacturing plant. This led to the creation of the Cadillac V-series for their luxury CTS sedan, sold as the CTS-V.

As with SUVs that have large-displacement engines, modern muscle cars are criticised for poor fuel efficiency. (The original muscle cars met with the same criticism in the 1960s.) However the muscle car is lighter at about 4000 lb (1800 kg) than the typical SUV, which weighs 4200 lb (1900 kg)-7200 lb (3300 kg).

Modern-day musclecars are subject to the gas-guzzler tax.

According to a 2006 press release, fuel economy standards forced GM to delay the Zeta platform when the Oshawa production facility had already been retooled for its production. The 2010 Chevrolet Camaro is one of GM's Zeta platform vehicles.

Australian Ford and Holden are currently producing high performance vehicles. For instance, Holden has its SS and SSV Commodores and Utilities, and HSV has more powerful Holden based versions and currently producing a limited edition HSV W427 - a Commodore fitted with the 7 litre V8 from the C6 Corvette Z06. Ford Performance Vehicles (FPV) turns out similarly uprated special versions of the Ford Falcon Sedan, the major difference being Ford offer a 360 hp (270 kW) turbocharged 4.0 litre I6 as well as their V8s. FPV are producing the GT 4-door Falcons—both Boss V8 and turbocharged sixes; the premier Fords are currently the BOSS V8 and F6 turbocharged inline 6.

Holden Special Vehicles currently produces high-performance versions of various rear-drive Holden Commodore sedans and, fitted with high performance (400 hp) V8 engines, and are perhaps one of the closest contemporary equivalents to the classic American muscle car (excluding the AWD of course)—fast, exciting, but relatively crude automobiles (though with far more attention to handling, suspension, safety and exceptional brakes compared with the stock models).

Vauxhall introduced the Monaro to the UK in 2004. This was a re-badged Holden Monaro fitted with a 5.7 litre Chevrolet Corvette engine, or in VXR form with the engine bored out to 6.0 litres. The car was reasonably priced and offered high performance for the pound and found a keen following, particularly after receiving ringing endorsements from television programmes such as BBC's 'Top Gear'. However sales were disappointing and the car was withdrawn from the Vauxhall range in 2007.

Collectibility

The original "tire-burning" cars such as the AMC AMX, Buick Gran Sport, Dodge Charger R/T, Ford Mustang, Oldsmobile 4-4-2, Plymouth GTX and Pontiac GTO, are "collector's items for classic car lovers."[44]

Surviving muscle models are now prized, and certain models carry prices to rival some of the more highly valued European sports cars. At auction the rarest vintage 1965–1972 muscle cars can be appraised at over US$500,000 depending on model, options, condition, demand and availability. Some rare models like the 1969 Chevrolet Camaro with the ZL1 option are considered the equivalent of real estate or museum relics.

Reproduction muscle-car sheet metal parts and even complete body shells are available.

Models

United States

Motor Trend identified the following models as "musclecars" in 1965:

- 1962–1965 Dodge Dart 413/426 Max Wedge/426 Hemi/Plymouth Fury 413/426 Max Wedge/426 Hemi
- 1964–1965 Ford Thunderbolt 427
- 1965–1969 Buick Skylark Gran Sport
- 1965–1970 Dodge Coronet/Plymouth Belvedere 426-S
- 1965 Chevrolet Chevelle Malibu SS
- 1965–1967 Oldsmobile Cutlass 442

Road & Track identified the following models as "musclecars" in 1965:

- 1964–1965 Pontiac Tempest Le Mans/GTO
- 1965–1975 Buick Riviera Gran Sport
- 1965–1969 Buick Skylark Gran Sport
- 1965–1970 Dodge Coronet/Plymouth Belvedere 426-S
- 1965 Chevrolet Chevelle Malibu SS
- 1965–1967 Oldsmobile 442|Oldsmobile Cutlass 442 [45]

Car and Driver also created a list of the 10 Best muscle cars for its January 1990 issue. The magazine focused on the engines and included:

- 1966–1967 Plymouth/→ Dodge intermediates with 426 Hemi
- 1968–1969 Plymouth/→ Dodge intermediates with 426 Hemi
- 1970–1971 Plymouth/→ Dodge intermediates with 426 Hemi
- 1966–1967 Chevy II SS327
- 1966–1969 Chevrolet Chevelle SS396
- 1968–1969 Chevy II Nova SS396
- 1969 Ford Torino Cobra 428
- 1969 Plymouth Road Runner/Dodge Super Bee 440 Six Pack
- 1970 Chevrolet Chevelle SS454
- 1969 Pontiac GTO

Other muscle cars include the following:

Muscle car

Full-size muscle models	Mid-size muscle models	Compact muscle models	Pony car muscle models	Muscle trucks
- 1961-1976 Chevrolet Impala SS until '69, then any high HP-engined models - 1958-1975 Chevrolet Bel Air - 1958-1972 Chevrolet Biscayne - 1965-1976 Chevrolet Caprice - 1959-1975 Ford Galaxie - 1959-1974 Mercury Monterey - 1960-1973 Dodge Polara	- 1970-1971 AMC Rebel and Matador The Machine[46][47] - 1968-1969 Buick Gran Sport - 1970-1974 Buick GSX - 1965-1973 Chevrolet Chevelle SS - 1966-1974 → Dodge Charger - 1968-1971 Dodge Super Bee - 1969 Dodge Charger Daytona[48] - 1966-1969 Ford Fairlane GT, GTA, and Cobra - 1968-1974 Ford Torino (GT, Cobra, and Talladega) - 1966-1972 Mercury Cyclone - 1970-1971 Mercury Montego - 1968-1971 Oldsmobile 442 - 1969 Oldsmobile Cutlass "Ram-Rod" 350 - 1970 Oldsmobile Cutlass W-31 - 1967-1971 Plymouth GTX - 1968-1974 Plymouth Road Runner[48] - 1970 Plymouth Superbird - 1964-1974 Pontiac GTO	- 1969 AMC SC/Rambler - 1971 AMC Hornet SC 360 - 1963-1974 Chevrolet Nova SS - 1968-1976 Dodge Dart GT, GTS, Swinger, and Demon - 1970-1976 Plymouth Duster - 1964-1969 Ford Falcon - 1970-1976 Ford Maverick Grabber - 1964-1975 Mercury Comet	- 1968-1970 AMC AMX - 1968-1974 AMC Javelin and AMX - 1967-1974 Chevrolet Camaro Z/28 & SS - 1970-1974 Dodge Challenger - 1965-1969 Shelby Mustang GT350 & GT500 - 1967-1971 Mustang Cobra Jet - 1969-1973 Mustang Mach 1 - 1969-1970 Boss 302 Mustang - 1969-1970 Mustang Boss 429 - 1971 Mustang Boss 351 - 1969-1970 Mercury Cougar Eliminator - 1964-1974 Plymouth Barracuda - 1967-1979 Pontiac Firebird & Trans Am	- 1965-1987 Chevrolet El Camino SS - 1967-1979 Ford Ranchero - 1971-1977 GMC Sprint - 1978-1987 GMC Caballero

Australia

Chrysler

VH model

- 1971-1972 Charger R/T E37 (101 built)
- 1971-1972 Charger R/T E38 - 280 bhp (210 kW) - 3 Speed Gearbox (Track pack and Big tank were options and a fully blueprinted engine) (316 built)
- 1972-1973 Charger R/T E48 (2 built)
- 1972-1973 Charger R/T E49 - 302 bhp (225 kW) - 4 Speed Gearbox (Track pack and Big tank were options and a fully blueprinted engine) (149 built)
- 1972-1973 Charger S/E E55 - 275 bhp (205 kW) - 727 Torqueflite Auto (340 cubic inch Chrysler LA engine) (124 built)
- 1969-1971 Valiant Hardtop (318 or 360ci V8s)

VJ model (R/T nomenclature dropped) were:

- 1973-1974 Charger E48 (169 built)
- 1973 Charger E49 (4 built)

- 1973-1974 Charger 770 E55 (212 built)

Ford
- 1967 XR Falcon GT (289)
- 1968 XT Falcon GT (302)
- 1969–1970 XW Falcon GT (351)
- 1969–1970 XW Falcon/Fairmont GS 302 and 351
- 1969 XW Falcon GTHO Phase I (351W)
- 1970 XW Falcon GTHO Phase II (351C)
- 1970-1971 XY Falcon/Fairmont GS 302 and 351
- 1970-1971 XY Falcon GT (351)
- 1971 XY Falcon Phase III GTHO (351)
- 1972 XA Falcon Phase IV GTHO 4 door (only four made: three prototypes, one production) (351)
- 1972–1973 XA Falcon GT hardtop coupe/4 Door Sedan (351)
- 1972–1973 XA Falcon GS Hardtop/Sedan/Ute (302, 351)
- 1973 XA Falcon Superbird (302)
- 1973–1976 XB Falcon GT hardtop coupe/4 Door Sedan (351)
- 1973–1976 XB Falcon/Fairmont GS Hardtop/Sedan/Ute (302, 351)
- 1974–1975 XB Falcon John Goss Special (302)
- 1976-1979 XC Fairmont GXL (302C or 351C as the desirable GT Power-pack Option)
- 1978 XC Falcon Cobra 5.8, Bathurst Homologation
- 1979 XD Fairmont Ghia ESP (302C, 351C)
- 1982-84 XE Fairmont Ghia ESP (302C, 351C)

Holden
- 1968–1969 HK Monaro GTS (327)
- 1969–1970 HT Monaro GTS (350)
- 1970–1971 HG Monaro GTS (350)
- 1971–1974 HQ Monaro GTS (350)
- 1974–1976 HJ Monaro GTS (308)
- 1970–1971 LC Torana GTR XU-1 (186)
- 1972–1973 LJ Torana GTR XU-1 (202)
- 1974–1976 LH Torana SL/R 5000 (308)
- 1974 LH Torana SL/R 5000 L34 (308)
- 1976–1978 LX Torana SL/R 5000 (308)
- 1976–1978 LX Torana SS (308)
- 1977 LX Torana SL/R 5000 A9X (308)
- 1977 LX Torana SS A9X (308)

Leyland
- P76 "Force Seven". This was a coupe version of the Leyland P76, and the company's answer to the Holden Monaro GTS, Ford Falcon GT and Chrysler Valiant Charger. The company ran into financial difficulties and ceased Australian production before the Force Seven could be released. The eight completed examples were sold at auction.

See also

- Pony car
- Personal luxury car

External links

Muscle Cars [49] at the Open Directory Project

References

[1] Koch, Jeff. "The First Muscle Car: Older Than You" Hemmings Muscle Machines - October 1, 2004 (http://www.hemmings.com/mus/stories/2004/10/01/hmn_feature17.html), retrieved on 2008-06-16.
[2] The Merriam-Webster definition is more limiting, "any of a group of American-made 2-door sports coupes with powerful engines designed for high-performance driving." car "muscle car." Merriam-Webster Online. Retrieved on 16 June 2008. (http://www.merriam-webster.com/dictionary/muscle)
[3] "Muscle Car Definition" Muscle Car Club Muscle, undated (http://www.musclecarclub.com/musclecars/general/musclecars-definition.shtml), retrieved on 2008-06-16.
[4] Sherman, Don. "Muscle Cars Now Worth Millions" The New York Times, June 4, 2006, (http://www.nytimes.com/2006/06/04/automobiles/04MILLION.html?_r=1&em&ex=1160020800&en=0e6a99c6df3961fe&ei=5087&oref=slogin) retrieved on 2008-06-16.
[5] Classic Muscle Cars Library, How Stuff Works, undated (http://musclecars.howstuffworks.com/classic-muscle-cars), retrieved on 2008-06-16.
[6] "Muscle Car Definition" by Muscle Car Society, undated (http://www.musclecarsociety.com/muscle-car-definition), retrieved on 2008-06-16.
[7] Henshaw, Peter (2004): *Muscle Cars*, Thunder Bay Press. ISBN 1-59223-303-1
[8] Mueller, Mike (1997). *Motor City Muscle: The High-Powered History of the American Muscle Car* (http://books.google.com/books?id=ZLP8kKL4w2kC&pg=PA13&dq=muscle+car+definition&ei=p-VWSLmdGI2ujAHm7IyYDA&sig=1gfGWDPXRhwQ0Eje4wDa7Cs1Z2g#PPA13,M1). MotorBooks/MBI Publishing Company. pp. 13. ISBN 978-0760301968. .
[9] "The Birth of Muscle Cars" by the auto editors of *Consumer Guide* (http://musclecars.howstuffworks.com/muscle-car-information/how-muscle-cars-work1.htm). Retrieved on June 03, 2008.
[10] Musclecars magazine, 1994.
[11] Nerad, Jack. "Oldsmobile Rocket 88", *Driving Today* (http://www.drivingtoday.com/kpix/greatest_cars/olds_rocket88/index.html).
[12] Dulcich, Steve: "Rocket Man" article in *Popular Hot Rodding* (http://www.popularhotrodding.com/features/0708phr_general_motors_432_ci_oldsmobile/index.html). Retrieved on June 07, 2008.
[13] Chrysler 300 article by the editors at Edmunds.com (http://www.edmunds.com/insideline/do/Features/articleId=102565). Retrieved on June 05, 2008.
[14] 1957-1960 "Rambler Rebel" by the auto editors of *Consumer Guide* (http://auto.howstuffworks.com/1957-1960-rambler-rebel2.htm). Retrieved on June 03, 2008.
[15] "AMC Muscle Cars" by the auto editors of *Consumer Guide* (http://musclecars.howstuffworks.com/muscle-car-information/amc-muscle-cars.htm). Retrieved on June 03, 2008.
[16] Shaw, Tom: "Anatomy of a Lightweight", *Legendary Ford magazine*, December 2005.
[17] "Ford Thunderbolt" article by the Auto Editors of *Consumer Guide*. (http://musclecars.howstuffworks.com/classic-muscle-cars/1964-ford-thunderbolt.htm) Retrieved on June 05, 2008.
[18] Holder, Bill, and Kunz, Phil (2006). *Extreme Muscle Cars*, Krause Publications. ISBN 0-89689-278-6.
[19] Gunnell, John (2005). American Cars of the 1960s, KP Books. ISBN 0-89689-131-3.
[20] "1966 Plymouth Satellite 426 Hemi" by the Auto Editors of *Consumer Guide*. (http://musclecars.howstuffworks.com/classic-muscle-cars/1966-plymouth-satellite-426-hemi.htm) Retrieved on June 12, 2008.
[21] "1969 Chevrolet Chevelle COPO" article by the Auto Editors of *Consumer Guide*. (http://musclecars.howstuffworks.com/classic-muscle-cars/1969-chevrolet-chevelle-copo-427.htm) Retrieved on June 05, 2008.
[22] "1970 Chevrolet Chevelle SS 454" by the Auto Editors of *Consumer Guide* (http://musclecars.howstuffworks.com/classic-muscle-cars/1970-chevrolet-chevelle-ss-454.htm). Retrieved on June 11, 2008.
[23] *Car Life* January 1969.
[24] "1968 Plymouth GTX" by the Auto Editors of *Consumer Guide* (http://musclecars.howstuffworks.com/classic-muscle-cars/1968-plymouth-gtx.htm). Retrieved on June 16, 2008.
[25] "1970 Plymouth Duster 340" by the Auto Editors of *Consumer Guide* (http://musclecars.howstuffworks.com/classic-muscle-cars/1970-plymouth-duster-340.htm). Retrieved on June 11, 2008.
[26] Kelly, Steve: "A new entry: DUSTER", *Hot Rod* March 1970.
[27] Genat, Robert (2006). *Mopar Muscle*, Motorbooks. ISBN 0-7603-2679-7.

[28] Cheetham, Craig (ed.) (2007). Ultimate Muscle Cars, Motorbooks. ISBN 0-7603-2834-X.
[29] 1970 "AMC Rebel Machine" article by the Auto editors of *Consumer Guide* (http://musclecars.howstuffworks.com/classic-muscle-cars/1970-amc-rebel-machine.htm). Retrieved on June 06, 2008.
[30] Kunz, Bruce. "1970 AMC Rebel", St. Louis Post-Dispatch, December 24, 2007 (http://www.stltoday.com/stltoday/autos/columnists.nsf/oldcarcolumn/story/E29F5B221D816D6F862573BB005A3428?OpenDocument). Retrieved on 2008-06-09.
[31] Nerad, Jack. "American Motors Rebel Machine", *Driving Today*. (http://www.drivingtoday.com/kpix/greatest_cars/amc_rebel_machine/index.html) Retrieved on July 01, 2008.
[32] Stunkard, Geoff. "Welcome To The Machine", *Mopar Muscle* magazine (http://www.moparmusclemagazine.com/featuredvehicles/mopp_0712_1970_american_motors_rebel_machine/index.html), retrieved on December 17, 2007.
[33] "Captain America: A Chopper Profile", by the Auto Editors of *Consumer Guide*. (http://auto.howstuffworks.com/captain-america-chopper-profile.htm) Retrieved on July 05, 2008.
[34] Kelly, Steve. "Too Much of a Rebel", *Hot Rod* magazine, February 1970.
[35] Sanders, Bill. *Motor Trend*, February 1969.
[36] "1968 Dodge Super Bee" by the Auto Editors of *Consumer Guide* (http://musclecars.howstuffworks.com/classic-muscle-cars/1968-dodge-super-bee.htm). Retrieved on June 11, 2008.
[37] Kelly, Steve. "Barracuda on the Line", *Hot Rod* magazine December 1968.
[38] Mel Nichols- HO down the Hume (http://falcongt.com.au/HO-DTH.html)
[39] quickest HDT according to *Modern Motor Magazine*, January 1983
[40] According to the Australian 5/2007 Wheels Magazine.
[41] "How Pontiac Works", by the auto editors of *Consumer Guide*. (http://auto.howstuffworks.com/pontiac32.htm) Retrieved on July 23 2008.
[42] Robinson, Aaron. "2008 Pontiac G8 GT – Road Test", *Car and Driver*, April 2008 (http://www.caranddriver.com/reviews/hot_lists/car_shopping/family_four_doors/2008_pontiac_g8_gt_road_test). Retrieved on July 24 2008.
[43] Robinson, Peter. "2006 Holden Commodore VE – Car News", *Car and Driver*, November 2006 (http://www.caranddriver.com/news/car_news/2006_holden_commodore_ve_car_news). Retrieved on July 24 2008.
[44] Zuehlke, Jeffrey (2007). *Classic Cars* (http://books.google.com/books?id=XI5-IBJ4bFAC&printsec=frontcover&source=gbs_summary_r&cad=0#PPT19,M1). Lerner Publications. pp. 18. ISBN 978-0822559269. .
[45] http://www.ratrodbarn.com/Rat-Rods-For-Sale/Muscle-Cars-For-Sale
[46] "1970 AMC Rebel Machine." by the Auto Editors of *Consumer Guide* 12 January 2007 (http://musclecars.howstuffworks.com/classic-muscle-cars/1970-amc-rebel-machine.htm), retrieved on 16 April 2009.
[47] Stakes, Eddie. "Matador Machine:1971 from American Motors" planethoustonamx (http://www.planethoustonamx.com/stuff/matador_machine.htm),, retrieved on 16 April 2009.
[48] Moriarty, Frank (1995). *Supercars: The Story of the Dodge Charger Daytona and Plymouth Superbird*. Howell Press. ISBN 978-1574270433.
[49] http://www.dmoz.org/Recreation/Autos/Enthusiasts/Muscle_Cars/

List of automobile model nameplates with a discontiguous timeline

- Alfa Romeo 8C (1931-1939, 2007-present)
- Alfa Romeo Giulietta (1954-1965, 1977–1985)
- Alfa Romeo Spider (1955–1993, 1995-date)
- Alfa Romeo SZ
- BMW 6 Series (1977-1989, 2003-present)
- Buick Century (1936-1942, 1954-1958, 1973-2005)
- Buick Roadmaster (1936-1958, 1991-1996)
- Buick Skylark (1953-1954, 1961-1972, 1975-1998)
- Chevrolet Camaro (1967-2002, 2008-present)
- Chevrolet El Camino
- Chevrolet Impala (1958-1985, 1994-1996, 2000-present)
- Chevrolet Malibu (1964-1983, 1997-present)
- Chevrolet Monte Carlo (1970-1988, 1995-2007)
- Chevrolet Nova
- Chrysler 300
- Chrysler Imperial
- Chrysler Town & Country
- Dodge Avenger (1995-2000, 2008-present)
- Dodge Challenger (2008-present)
- → Dodge Charger (1966-1978, 1983-1987, 2006-present)
- Dodge Magnum
- Dodge Monaco
- Ferrari Mondial (1954, 1980-1993)
- Ferrari Testarossa (1956-1961, 1984-1991)
- Ferrari California (1960, 2008-present)
- Fiat 500 (1937-1955, 1957—1975, 1991-1998, 2007-present)
- Fiat 600 (1955-1969, 1998-present)
- Fiat Bravo (1995-2002), 2007-present)
- Fiat Croma (1985-1996), 2005-present)
- Ford Thunderbird (1955-1997, 2002-2005)
- Holden Astra (1984-1989, 1995-current)
- Holden Monaro (1968-1976, 2001-2006)
- Holden Statesman (1971-1984, 1990-current)
- Holden Ute (1951–1985, 1990–current)
- Kia Sportage
- Lincoln Continental
- Lincoln Zephyr (1936, 2006)
- Mercury Capri
- Mercury Cougar
- Mercury Montego
- Mercury Monterey
- Mercury Sable (1986-2005, 2008-present)
- Mitsubishi GTO (1970–1975, 1990-2001 without the Galant nameplate)

- Mitsubishi FTO (1971–1974, 1994–2000 without the Galant nameplate)
- Nissan 200SX (USA market only)
- Nissan Silvia (1964–1968, 1974–2002)
- Nissan GT-R (1969-1973, 1989-2002, 2008-present; without the Skyline nameplate)
- Nissan Z-car (1969-2000, 2002-date)
- Nissan Fairlady (1960-2000, 2002-date)
- Porsche Carrera GT (1980-1982, 2006)
- Porsche Spyder
- Porsche Speedster (1954-1962, 1987-1993)
- Plymouth Duster
- Pontiac LeMans
- Pontiac GTO
- Pontiac Tempest
- Škoda Felicia (1959-1963, 1994-2001)
- Škoda Octavia (1959-1971, 1996-present)
- Škoda Superb (1934-1942, 2001-present)
- Toyota FJ (1953-1984, 2006-date)
- Volkswagen Beetle/Volkswagen New Beetle (see notes)
- Volkswagen Rabbit

Notes

- Skipping model years 1943 through 1945 due to World War II is notwithstanding for the purposes of this list.
- Volkswagen Beetle/New Beetle: discontinuous in major markets, though the "legacy" Beetle continued production for sale in Mexico.
- The Mercury Monterey was reincarnated as a minivan.
- Chrysler Town and Country was reincarnated as a station wagon in the 1980s, then as a minivan since 1990.
- Nissan's Z-car was discontinued in 2000, and resumed production with the 350Z.
- The GT-R continued as a model separate to the Skyline
- The first Porsche Carrera GT was based on the 356 in the early 1960's, the second one was a LeMans Homologation model based on the 924, and the third Carrera GT was a bespoke supercar.
- The Porsche Spider was reincarnated as an LMP2 Race Car
- The Chevrolet Corvette might be misconstrued as discontiguous, due to the 1983 model year. As there were problems with the new model (C4), all but a couple of the 1983 vehicles were destroyed. The survivors currently reside in museums, especially the Corvette Museum in Bowling Green, Ky, out of private hands.

Article Sources and Contributors

Dodge Charger (B-body) *Source*: http://en.wikipedia.org/w/index.php?title=Dodge_Charger_%28B-body%29 *Contributors*: 293.xx.xxx.xx, Abramsgavin, Aldis90, Andy Christ, ApolloBoy, Aspects, Atom davis, AuburnPilot, Beyond the Grave, Bill Wrigley, Black arrow, Boxofrox50, Bugo, Bull-Doser, CL, CZmarlin, Caesar Rodney, Caknuck, Cargocontainer, Chenzw, Chris the speller, Ckatz, Cmcfarland, ColdShine, Colonies Chris, D.valued, DSRH, Daemon8666, Dbergan, DeadEyeArrow, DerHexer, Discospinster, Doc Tropics, Dodge Don, Dodgegirl19, Donnie Park, EJF, ENeville, East718, Eubulides, FastbackJon, Finngall, Fraggle81, Frankie0607, Gert527, Gilliam, Go229, Gonzo fan2007, HDCase, Hailey C. Shannon, Hawk91, Henrik, Howcheng, Hu12, Husond, IFCAR, Interiot, Iridescent, Irishguy, J.delanoy, Jimpatnmatt, Jonny-mt, KLRMNKY, Karrmann, Kelly Martin, Kevin Breitenstein, Klow, Lady Aleena, LanceBarber, Lavenderbunny, Littleman TAMU, MER-C, MJBurrage, Manway, Martyn 2 Martin, Mattditty, Mboverload, Mentifisto, Mephistau, Mikthomp, Moncrief, Moogyboy, Moto100, Mpotter, Mrand, Mrceleb2007, Nepenthes, Nick, Nv8200p, Ohnoitsjamie, Omicronpersei8, Orlin Kvasir, PS2pcGAMER, Pegship, Peteywick, Poeloq, Prkr 07, Radon210, Rettetast, Ridge Runner, Sable232, Santaclause157, Schmendrick, Sebasstian, Sfoskett, Silent SAM, Sixtyeight, Somno, Strannickxxx, Stwalkerster, Superbeefinder, Tabletop, Thatcher, TheDJ, Thunderstrike66, Tom.k, Trevor Wennblom, Typ932, Undead warrior, Upex, Utcursch, UtherSRG, WOSlinker, Wallygarza, Wasted Time R, Whitelotusclan, Wikiuser100, Willirennen, Wmahan, Xi311, Zippy, Zoicon5, Zubdub, 410 anonymous edits

Dodge Charger *Source*: http://en.wikipedia.org/w/index.php?title=Dodge_Charger *Contributors*: Alansohn, ApolloBoy, Barkeep, Bkonrad, BrendelSignature, CZmarlin, Calvin 1998, Camw, Castletower, Ckatz, Cmdrjameson, Darth Panda, Deichkind, DerHexer, Dodgegirl19, Dr Dec, Everyking, Ezhiki, Ezn, F. Cosoleto, FastbackJon, Figureskatingfan, Gert527, Gordonfan, Hailey C. Shannon, Hellbus, Hu12, IRT.BMT.IND, Jacek Kendysz, Jeffrey Mall, Jennifer.craft, Johng, Josh Parris, Justinc, KBBEditors, KLRMNKY, Kate, Leedeth, Lommer, Magicmonster, Maximus Rex, Meegs, Morven, Mphmag, MrDolomite, Na2rboy, Nakon, NiteowlneiIs, NormanEinstein, Nrbelex, Nsaa, Num1dgen, Oilpanhands, OlEnglish, Pearle, Pradico, R'n'B, Radagast, Rascalb, Rebug, Ridge Runner, RivGuySC, Rnkibbe, Roadstaa, Rsduhamel, RxS, SaltyPig, Schmendrick, SchmuckyTheCat, Scottfisher, Sfoskett, Shadowjams, Stombs, SuperDude115, Superbeefinder, Tenev, The Thing That Should Not Be, Tide rolls, Tohd8BohaithuGh1, Traveler100, Trekphiler, Tyler dowell, Typ932, Unschool, Versus22, Xen 1986, Zeromaru, Zubdub, 185 anonymous edits

Car model *Source*: http://en.wikipedia.org/w/index.php?title=Car_model *Contributors*: Aldo L, Biscuittin, Chris the speller, Enzofrolan, Gaius Cornelius, Gzuckier, Ham Pastrami, Heds, Ibrahim89, Interiot, Keith D, Kierant, Mac, Malcolma, NaBUru38, Neoking, OSX, Radagast83, Shimgray, Snigbrook, Spacepotato, Usa1, Ts, 37 anonymous edits

Automobile *Source*: http://en.wikipedia.org/w/index.php?title=Automobile *Contributors*: $5forMe, -Midorihana-, -hiphop-, .K, 04pandre, 08toi, 10sh10, 10tottl10, 200.191.188.xxx, 21655, 24.93.53.xxx, 2586, 271292hrs, 2e2e2e3r3r4t4tt5y5y6u6u7ui8i, 80120t, 83d40m, 92889, AEMoreira042281, AGToth, AKMask, ATS 500, Aaron Brenneman, Abdullais4u, Abkovalenko, Academic Challenger, Acdx, Adam Rock, Adam78, Adamkelly, Adashiel, Ahildaballa, Ahoerstemeier, Aim Here, Akadruid, Aksi great, Ala.foum, Alai, Alan Liefting, Aldie, AlexPlank, AlexWilkes, Alexdi, AlexiusHoratius, Alexofdodd, Alexzor1, Alfio, Ali K, Aliceinlampyland, Alison, Alison9, All Is One, All in, Alskdjfdhfiwow, Altenmann, Altermike, Ambarish, AmosWolfe, Anand Karia, Ancheta Wis, AndonicO, Andre Engels, Andrea244, Andrewferrier, Andrewpmk, Andrewrhchen, Andy G, Andy M. Wang, AndySimpson, Andycartland, Andycjp, Andygharvey, Anetode, AniRaptor2001, Animum, Anobo, Anonymous Dissident, Anonymous anonymous, Anphanax, Antandrus, Antonrojo, AntonyGDines, Anwar Javed, Anwar saadat, ApolloBoy, Archangel x105, Aremith, Argon233, ArielGold, Arpingstone, Arrenlex, Arsenikk, Asanchez1572, Asfadsfasdfasdfasdfreybhkiu, Ashenai, Asskikr1001, AtholM, Atmoz, Atn55, Aude, Autonova, Avarame, Avengerx, Average Earthman, Az1568, Aztune, BECPL, BGOATDoughnut, BRG, Barneca, Barnie slyvester, Barrylb, Basawala, Bathrobe, Beckette, BeelzebubJNR, Beetstra, Beland, Belgian man, Bencherlite, Benefactordyr, Benny the wayfarer, Beno1000, Benzocane, Berney, Berria, Bevo, Bewildebeast, Beyond silence, Bfar7122, Bfigura's puppy, Bggoldie, Bhadani, Bhludzin, Bilbobee, BillC, BillFlis, Birdthug, BiroZombie, Biswaranjan, Bkonrad, Blablablob, Blah3, Blainster, Blank7, BlankVerse, Blimpguy, Bloodshedder, Blorf, Blue520, BlueMars, Bntsg, Bob332, Bobblewik, Bobet, Bobjob455, Bobo192, Bobobo10547, Bobobobobobobobobobobobobobobobobobobobobobobobob, Bogdangiusca, Boivie, Bonaparte, BonesBrigade, Bongle, Bongohead222, Bongwarrior, Boothy443, BorgQueen, Borisblue, BostonMA, Bovineone, Brakedust, Branden Gorecki, Brat32, Bravada, BrendelSignature, Brian Pearson, Brian0918, Brim, BrokenSphere, Brossow, BrowardBulldawg, Bruce A. McHenry, BryanG, Btownsox13, Buckley699, Burzmali, Byrule62, CART fan, CIMCOBILL, CJ, CSvBibra, CUSENZA Mario, CZmarlin, Cacycle, Cadillac, Calm, Caltas, Camaro96, Camaron, Cambrasa, CambridgeBayWeather, Camembert, Cameron Dewe, Can't sleep, clown will eat me, Canderson7, Candyman36ws, Captain panda, CardinalDan, Careless hx, Carfreek, Carmaniacx, Carmenharvey, Casey8572, Caster23, Cate, Catmoongirl, Celebration1981, Century0, Ceyockey, Chan Han Xiang, Chaos, Charitwo, Charles Matthews, Charmander trainer, Cheese wizzz, Chendy, Chief tin cloud, Chill doubt, Chouette Sage, Chris 73, Chris Roy, ChrisUK, Chrisd87, Christian List, Christopher Parham, Chu333222, Chupa me wevos, CiTrusD, Clampton, Cleanupman, Clearbluepr, Cliché Online, Cobi, Colleenthegreat, Cometstyles, Compass35, Conufsed, Conversion script, Cookie90, Cool Blue, Coolcaesar, CopperMurdoch, Cory007, Covalent, Cowboy Ray, Craig Butz, Crazycomputers, Crystallina, Curley boi 69, Cuye, CyclePat, DAMurphy, DH85868993, DJ Bungi, DJ Clayworth, DRTllbrg, DVD R W, Da man239, Da monster under your bed, DanKeshet, DanMS, Daniel, Danmore1101, Dar-Ape, DarkFalls, Darkfred, Dave Runger, David R. Ingham, DavidLevinson, DavidMcKenzie, DavidWBrooks, Davidgoymt, Davidovarcherson, Dbergan, Dbtfz, Dcoetzee, Ddgonzal, Ddmtrackstar, DeFacto, DeLarge, DeadEyeArrow, Delirium, Delldot, Deltabeignet, Deor, DerHexer, Derek Ross, Descendall, Dharmabum420, Dietary Fiber, DifiCa, Digitalme, Digitize, DineshAdv, Dino246, DirkvdM, Dismas, Dispenser, Djinn112, Djordjes, Dlae, Dman64, DocWatson42, Doctorevil64, Doctormatt, Dodo bird, Dorfd1, Dorftrottel, Doug Coldwell, DougieLawson, Dr demon, DrBob, Dragonflyharasser, Drini, Dryazan, Duffman, Duk, Dust Filter, Dysepsion, ESkog, ETO Buff, Eco84, Ed g2s, EdJogg, Edcolins, Eddy3615, Edward, Edward321, Egil, Ekrub-ntyh, ElTyrant, Elenseel, Eliz81, Elmicker, Eluchil404, EmanWilm, Emarshizz, Emc2, Emre D., EnSamulili, Enigmaman, Enthusiast10, Envirocorrector, Erc, Erebus555, Eric outdoors, EricStyles, Eried, Esprit15d, Eternalnoob, Etiquette, Eubulides, Eukesh, Evankk, Everyking, Evil Monkey, Evil saltine, Exander, Excaliburo, Exia, Eye of the Mind, EyrianAtWork, Eyu100, Fabricationary, Fact check, Fang Aili, Fanoftheworld, Fantadrink11, Fdp, FelisLeo, FenderTele, Ferkelparade, Ferracin, FetchcommsAWB, Fir0002, Firefoxman, Firsfron, FishUtah, FisherQueen, Fixa, Flonase, Flyerhell, Flyguy649, Foxandpotatoes, Foxman95, FrancoGG, Frank, Frankn628, Freakofnurture, Fredbauder, Fredrik, Freeke, Freiberg, FreplySpang, Freshbakedpie, Frymaster, FuegoFish, Funnybunny, Fvw, Fwappler, G Clark, GDonato, GPrestidge, GRAHAMUK, Galadh, Gamaliel, GangstaEB, Gansys, Gardar Rurak, GargoyleMT, Gary King, Gary99, Gciriani, Geil ramirez, Gene Nygaard, Geneb1955, Geologyguy, George737, GeorgeMoney, Ghaly, Ghettodude, Ghewgill, Ghostbubbles, Ghostshark, Giftlite, Gigemag76, Gilliam, Gjd001, Glen, Glenn, Globeclotter, God of War, Gogo Dodo, Gogoboots, Goodm08, Graemel., Grafikm fr, Graham87, Graufbold, Gravitan, Green caterpillar, Greennecro8, Greensidedown, Greentopia, Gregfitzy, Greglocock, Greyhood, Ground Zero, Gtapro91, Guaka, Gurch, Gurlpower Ca, Guska, Gwernol, Hadal, HaeB, Happymango, Happynoodleboycey, Harold1234, Harryt95, Hbackman, Hbird18, Hehehe111, HeikoEvermann, Hektor, Henry Flower, Henry1245, Henrygb, Hermione1980, Heron, Heymrdrummer, Hezzy, Hgvlyiwegv, Hi.ro, Highwaydude2007, Hiper15, Historikos, Hoohp, HorsePunchKid, Hotlorp, Hoxxy, Hu12, Hullo exclamation mark, Husainj, Hut 8.5, HybridBoy, I Dig Cars, IANMOORELOVESWIKI, IGod, IRT.BMT.IND, Iain99, IamSuzie, Iamvered, Ian Dalziel, Icairns, Icarus, Icewedge, Icot, Icydid, Igor potkonjak, Ihosvany, Ikeses, Ikh, InNuce, InfinityAndBeyond, Infrogmation, Inter, Interiot, Ireas, IronDuke, Ixfd64, J 1982, J Di, J. Finkelstein, J.delanoy, JForget, JLaTondre, JRM, JTN, JYolkowski, Jaberwocky6669, JackLumber, Jackace911, Jaganath, Jak Inn Thee Been Stock, Jakey981, Jakslayer, James086, JamesBurns, Jasonstark, Jasz, Jauerback, Java7837, Jayron32, Jdforrester, JeLuF, JeopardyTempest, Jeremyster, Jeromewiley, Jespersm, Jesse Viviano, JesseGarrett, Jesvane, Jh51681, Jimbobzeway, Jimp, Jimxu, Jjron, Jmlk17, Jnc, JoanneB, Jobe457, Jodis, Joe11miles, Joeblakesley, John254, JohnCub, Johnie, Johnmccollim, Johnny69420, Jon442, Jonearles, Jonnogibbo, Jononon, Jorge Stolfi, Jose77, Joshthejedi3, Joshygoodwin, Jossi, Joy, Joyous!, Jpgordon, Jph, Jpogi, Jptwo, Jredmond, Jtdirl, Jtkiefer, Jules.lt, Juliancolton, JuliusJackson, Junglecat, JunkyardDawg, Just zis Guy, you know?, Justinlovesmonica, Jwhyte, Jwy, JzG, K731, KBBEditors, KFan II, KJBracey, KVDP, Kageyose, Kaisershatner, Kanags, Kane5187, Kar98, Karrmann, Katalaveno, Kateshortforbob, Katieh5584, Kazrak, Kazvorpal, Kdales, KelleyCook, Ken6en, Kendrick7, Kesac, Kevin, Kevin Hayes, Kevyn, Keycard, Kiand, Kidpoker15, Kierant, Kim Bruning, Kingturtle, Kipala, Kiteinthewind, Kjkolb, KnowledgeOfSelf, Kntrabssi, Konman72, Korath, Kostan1, Kowey, Krawi, Krishowns, Krsont, Kukini, Kuribosshoe, KurtRaschke, Kuru, Kurykh, Kuzaar, Kvdveer, Kyle sb, Laogeodritt, Lapuwali, Latka, LaulLaunohow, Lda523287, Lectonar, Ledpup, Leenewton, Legotech, Leidolf, Leonard G., Lesorcan, Levineps, Lexor, Liftarn, Lightdarkness, Lightmouse, Like 3, Limp10, Linmhall, Llort, Llywrch, Locos epraix, Logan, LonelyMarble, Lord Hawk, Lord Pistachio, Lord Ramza, Louisrix, Lovenoble, Luapnampahc, Lucan8ter, Lucas the scot, Lucasbfr, Lucasgarsha, Lucyin, Lugomasa, Luk, Lumos3, Luna Santin, Lupin, Lupo, LuvWikis, Lynndunn, M1ss1ontomars2k4, MD87, MER-C, MZMcBride, Mac, MacMed, Magain, Mailer diablo, Mairi, Malcolma, Mani1, Maniacgeorge, Marc Venot, Marcus22, Mariordo, Mark Dingemanse, MarkGallagher, Markb, Master of Puppets, Mateuszica, Mateycow, Matron10, Matt Crypto, Mav, Maximus Rex, Mayank shah 89, Mcla11pt1, Mcluvin, Mcmillin24, Mdlutz, Meeples, Meggar, Mel Etitis, Mellisa Anthony Jones, Mentor2007, Mercury888, Merphant, Meyers018, Mgreenbe, Michael Hardy, MickMacNee, Midgrid, Midnightcomm, Miguelisgay, Mike 7, Mike Christie, Mike Rosoft, MikeCapone, Mikem77, Mikeymcbreen, Milkmandan, Million$Baby, Mindspillage, Minesweeper, Minority Report, Mintguy, Mion, Miss Madeline, Misza13, Mjb, Mjpieters, Mlouns, Modster, Monedula, Moneypage, Montrealais, Moonriddengirl, Moreschi, Moros, Morther, Morven, Moscovite Knight, Mr Mulliner, Mr. Lefty, Mr.hankythepoo, MrDarcy, MrDarkChao, Mrc360, Msikma, Muchness, Mulad, Muppppet, Murderbike, Mushroom, Musical Linguist, Mxn, Myanw, Mysdaao, Myths245, NaBUru38, Nakon, Naryathegreat, Natofyelsew, NawlinWiki, Nayvik, Neil Martin India, Neonumbers, Nerval, Netsnipe, Nevetsjc, New Thought, NewEnglandYankee, Newfoundlanddog, Ngantengyuen, Nighthawk leader, Nikkul, Nil Einne, Nilfanion, Nilmerg, Ninnnu, Nishkid64, NiteowlneiIs, Nivix, Nju987, Nk, Nkayesmith, Nlu, Noah Salzman, Node ue, Non-dropframe, Nonono123, Nopetro, Norm, Nslsmith, NuclearWarfare, Nucleusboy, Nufy8, Nukeless, Nyletak, Nyvhek, ONEder Boy, OSX, Obli, Ocicatmuseum, Ohnoitsjamie, Oilpanhands, Oleg Alexandrov, Olegwiki, Oliver202, Olivier, Omgwehavename, Omicronpersei8, Onevalefan, Ontopic, Orangutan, Ott, OverlordQ, OwenX, Oxymoron83, P0mbal, PAK Man, Paco1991, Padajtsch-kall, Palica, Panchitaville, Papaya07, Parable1991, Past-man, Patrick, Patstuart, Paul August, Pavel Vozenilek, Pawyilee, Pax:Vobiscum, Paxsimius, Pc13, PeaceNT, Peas and corn, Pediainsight, Peligroso, Pellande12, Persian Poet Gal, Peruvianllama, PeteWailes, Peter, Peter Horn, Pgk, Phill up 007, Phlamer26, Phntm, Pigeon.dyndns.org, Pjbflynn, Pladuk, PlatinumX, Pokrajac, Polluxian, Pomadeguy, Poorleno, Porqin, Portiab, Possum, Poypoyboiiii, Prashanthns, Primate, PseudoSudo, Ptk, Puchu, Pumbill, Punchup, Pwnr77, Pyrospirit, Q5967, QIOY, Qbmessiah, Qqwwwwwwww, Qtoktok, Quadell, Quadratic, Quaque, Quiddity, Qwerty33, Qxz, R9tgokunks, RC-0722, RG2, RJN, RMFan1, RWardy, Rachelgiggles, RadioKirk, Radon210, Randomblue, Raryel, Rasmus Faber, Razbox, Rdsmith4, Recury, Red Thunder, RedCoat1510, Redvers, Reginmund, Regushee, Relvin, Remag Kee, Renamed user 24, Rent A Troop, Reswobslc, Rettetast, Revcanon, RexNL, Rgoodermote, Rhobite, Riana, Ric36, Rich Farmbrough, Richardearl, RickK, Rintrah, RitKill, Rjwilmsi, Rlsheehan, Rmhermen, Rmt2m, Roadrunner, Rob230, RobDe68, Robert Merkel, RobertG, Robitaille03, Rocastelo, Rock427, RockMFR, Rocket000, Rodrigue, Roke, Rolf, Romanm, Rory096, Roux, RoyBoy, Royalguard11, Rujoesmith, Runch, Runner5k, Rusl, Ryan4314, Rzf3, S.M., SCEhardt, SPUI, Saberwyn, Sagesebas, Saket6, Sam Francis, Sam Hocevar, SamH, SamdupLama, SamuraiClinton, Samw, SavoySison, Sc147, Sceptre, Schector, Schicklgruber, Schlichtinator, SchmuckyTheCat, Schwnj, Scientizzle, Scipius, Sdornan, Sean Whitton, Sean William, Seant, Selimbey, SenorAnderson, Seoanalyzer, Sfoskett, Shacurthi, Shadow1, Shane0016, Shanes, Sharnak, Shashwatrocks, Shawnc, Shazam3592, Shenme, Shoemoney2night, Siddhant, Sietse Snel, Silverback, Silverxxx, SimonP, Sir Lewk, SirVulture, Sjakkalle, Sjschen, Sk8er 4533, SkiBumMSP, Skull365, Sladen, Slakr, Slippered sleep, Smidydevil, Smith609, Smmf1, Snailwalker, Snori, Snowmanradio, Snoyes, Soakologist, Sobolewski, Sohailstyle, Sohmc, Solipsist, Soriamic, Souljaboytellem44, South Bay, Sp, Spangineer, Spearhead, Specter01010, Speeddemonvegas, Spellcast, Spick And Span, Spike Wilbury, Splintax, Spookfish, SpookyMulder,

SportWagon, Springeragh, Sqash, Staceybabesx, Stagerd.thomas, Stalfur, Stapian, Stars4change, Steel, Steinsky, Stephan Leeds, StephanGFX, StephenMacmanus, Stephenb, SteveBaker, Steven Zhang, Stewacide, Stickman sam, Stombs, Stoopered, Stuartclift, Stude62, StuffOfInterest, Sunroof, Super edd, SuperDude115, Superkevin95, Swift, Syrthiss, Syvanen, TACD, TBadger, Talking goat, Tannin, Tasc, TastyCakes, Tbjablin, Tbonnie, Tcotrel, Teddey, Teilas, Teklund, Tellyaddict, Terps21, Terra-rent, Terrorist420x, Teryx, Teutonic Tamer, TexasAndroid, Thalia42, The Evil Spartan, The Giant Puffin, The Ronin, The Transhumanist, The undertow, TheBlazikenMaster, TheKMan, TheNewPhobia, Thelacerator, Themanwithoutapast, Theonlycarmire, Theuion, Thewalrus, Thingg, Tiddly Tom, Tifego, Tiptoety, Titoxd, Tkeller62, Tolu600, Tom harrison, Tombstone, Tomdobb, Tone, Topory, Toppersjcc, Train of thought, Trekky3012, Trekphiler, Trent, Trevor MacInnis, TrippingTroubadour, Tristarmotors, Trozig, Trusilver, Truth or consequences-2, Tstingert, Tsunaminoai, Ttwaring, Tucson Arizona Mexico, Typ932, UNIXCOFFEE928, Ultratomio, Ultraviolet scissor flame, Ummakynes, Ummit, Uncle G, UniQue tree, User27091, Vaceituno, Val42, Valuecityauto, Van helsing, Velho, VeryVerily, Vgranucci, Vildricianus, Voice of All, Vsbhogadi, W4zz0, WJBscribe, WODUP, Waggers, Waldopepper, WalrusRescue, Walton One, Watti Renew, Wavelength, Wayward, Wetwilli, Whale plane, WhaleyTim, Whiteshoes, Whomp, Wickethewok, Wik, Wiki alf, Wikibofh, Wikidudeman, Wikiklrsc, Wiktionary4Prez!, Wilt, Wimt, Winchelsea, Wingspeed, Wipe, Wknight94, Wmahan, Wolfkeeper, WongFeiHung, World arm lamp, X3ni, Xagent86, Xcentaur, Xen 1986, Xerocs, Xezbeth, Xkfusionxk, Xymmax, Yamaguchi先生, Yankees9043, Yazeed M, YellowMonkey, Yelsent, Yidisheryid, Yihfeng, Yitscar, Ylbissop, Ylem, Ynhockey, Yoasif, Youngoat, Youpli, Your vexation, Z.E.R.O., ZS, Zafiroblue05, Zazou, Zeno Gantner, Zeno of Elea, Zepheus, Zigger, Zmajjlu, Zombiebaron, Zondor, Zoney, Zsinj, Zzuuzz, Tג, ליבי לב, 2389 anonymous edits

Dodge *Source*: http://en.wikipedia.org/w/index.php?title=Dodge *Contributors*: 21655, 23ph98apfrqz3h2, 68DANNY2, 842U, 8mile4, A Second Man in Motion, AJ-India, Acela Express, Acsenray, Adamlane, Ahoerstemeier, Akadruid, Alansohn, Andyroo31, Angela2109, AnnaFrance, Antonio Lopez, ApolloBoy, Arj26, ArmadilloFromHell, Arthena, Assassinoc714, Atn55, Avalanche08, BRG, BSI, Bassbonerocks, Bennygubear, BetaMaxx11, Bewenched, Bobo192, Bongwarrior, BrendelSignature, Brewcrewer, Brian in denver, Brizzleness, Brossow, Brutaldeluxe, Bsadowski1, Bull-Doser, BulldozerD11, Bungofpot, Burgundavia, Buron444, Bytwerk, CZmarlin, Can't sleep, clown will eat me, Canadian-Bacon, Carom, Carpictures, Caz-kenny, Choalbaton, Chowbok, Chrycoman, Chryslercom, Ckape, Ckatz, Clawed, Closedmouth, Collard, CoolKid1993, Crevox, Cryptic, DBCaravan, DanTD, Danieldanieldaniel12345, DavidWBrooks, Dawn Bard, Dbfirs, DeLarge, Desmay, DiRF, DigitalC, Digitalme, Discospinster, Dma124, DocWatson42, Doodsweat, DoubleBlue, DougW, DowneyOcean, Dreadstar, Drilnoth, DropDeadGorgias, Drussel3, Dtobias, Duncan1800, Duncharris, Dyno Tested, Dysepsion, EarlevilleRedneck, Either way, Ejfetters, ElSaxo, Elcobbola, Elockid, Epbr123, Equendil, Erik9, ErikNY, Errorlines, Etienne, FastbackJon, Feitclub, Fieldday-sunday, Fluffysheep15, FogDevil, Frankenpuppy, Frantik, Fratrep, Garciaegs, Garion96, Gauchocxj, Gidonb, Gilliam, Glane23, Goodnight565, Goulais69, GraemeL, GreaterWisdom, GreatestWisdom, Gruzd, Gurch, Haemo, Hall Monitor, Harald Hansen, Heegoop, Hemi V8, Hobartimus, Home Suggestion, Hooperbloob, Hrajt, Hu, Hu12, Hut 8.5, Il MusLiM HyBRiD II, IRT.BMT.IND, Ianris, Igoregor007, ImGz, Importmyvehicle, Infrogmation, Iridescent, Irishguy, J.delanoy, JHMM13, JaGa, Jacob Poon, Jamie12321, Janus657, JarlaxleArtemis, Jaydeah, Jesquardo, JePe, Jefflundberg, Jengod, JeremiahZ, Jesandor, Jetlag, Jimpatnmatt, Jj137, John Anderson, Jollyroger, Jtrevan, KansasCity, Kariteh, Karrier, Karmmann, Kbdank71, Kierant, Kitopher2001, Kotra, Kozuch, KsprayDad, Kwsn, LanceBarber, Landroo, Leonard G., Lglswe, Lightdarkness, Linkboyz, Lost on belmont, Lotus L-12, Lsb43, LugPaj, Luna Santin, Lurkster, Lyght, MacX1993M, Maddie122, Magog the Ogre, Malcolma, Marknew, Martarius, Maser Fletcher, MattKeegan, Mav, McSly, MediaRocker, Megassgohan5, Mentifisto, Mesto, Milkmandan, Millisits, Miquonranger03, Mmsimpson, MoparManiac89, Moriori, Morven, Mpotter, MrDolomite, Mrclassicauto, Mrleechamberlain, Mtnerd, Mufka, Mygerardromance, NRAPA33, NaBUru38, Nakon, NellieBly, Nevercloud, NewEnglandYankee, NickMartin, Nicko90161, NiteowlneiIs, Norm, NormanEinstein, Nyj1218, Oda Mari, Ohnoitsjamie, Omicronpersei8, Oneforshodak, Onorem, Opertinicy, Orangemarlin, OwenX, Pc13, Philip Trueman, PhoenixMourning, PrinceGloria, Prolog, Proski, PurplePower46, Pwt898, Pyfan, Qero, RadicalBender, Rairdon, Randroide, Rascalb, Recury, Redd Dragon, Redvaiz, Res2216firestar, RexNL, Rgamble, Rj, Rjd0060, Rjwilmsi, Robbie Cook, Rodrigo123456, Ronald20, Rsnetto74, Rsrikanth05, SRT8, ST47, Sable232, Sceptre, Scheinwerfermann, Scientizzle, Sfoskett, Sguler, Shadowrunner22824, Shhh90, Silivrenion, Sinneed, Sjsitler92, Skittleys, Slo-mo, Smithc 321, Snowolf, SpaceFlight89, Speedyduner, Squash Racket, StaticGull, Stephen, Stephenplodinec, Stude62, Subman758, Supadawg, Superbeefinder, Swaq, T-dot, Teapotgeorge, Ted Wilkes, Texture, The Flying Spaghetti Monster, The Helper S, TheKMan, Thingg, Thomas Paine1776, Tide rolls, Tillman, Tinton5, Tiptoety, Tms., Tomahawk101, Tregoweth, Trekphiler, Triddle, TubularWorld, Turk oğlan, Typ932, Ugur Basak, Ukexpat, Vegavairbob, Vfw post 7426, Villager57, Vossanova, W00d, Waninge, Wars, Warut, WideArc, Wik, Wiki alf, Wonk619, Woohookitty, Xljesus, Xnatedawgx, Xy7, Zer3k, Zoe, 660 anonymous edits

Chrysler B platform *Source*: http://en.wikipedia.org/w/index.php?title=Chrysler_B_platform *Contributors*: ApolloBoy, BeepBeepRoadrunner, Brossow, ENeville, Either way, Joeychgo, MSJapan, Rich Farmbrough, Sales2007, SaltyPig, Sfoskett, Zoicon5, 4 anonymous edits

Chrysler Cordoba *Source*: http://en.wikipedia.org/w/index.php?title=Chrysler_Cordoba *Contributors*: 93JC, Almondwine, ApolloBoy, Bavaria II, Boysrch, Bull-Doser, CZmarlin, Chivista, Chrislk02, CommonsDelinker, Contributor168, DLemasa, DanMS, Dominic, Donnie Park, Dystopos, Either way, Fuhghettaboutit, Gaius Cornelius, Gloriamarie, Golfman, IFCAR, J newkirk, Jnelson09, Jon the dodgeboy, Kingclyde, Kinu, MSJapan, Morven, Norm mit, Nv8200p, O^O, Reezy, RivGuySC, Rlandmann, ST47, Sir Nicholas de Mimsy-Porpington, The Fat Man Who Never Came Back, Tooslim73, Typ932, Vossanova, Voyonyx, White 720, 49 anonymous edits

Chrysler *Source*: http://en.wikipedia.org/w/index.php?title=Chrysler *Contributors*: 1122334455, 163.200.81.xxx, 16@r, 68DANNY2, 842U, A Second Man in Motion, AJ-India, AbsolutDan, Acela Express, Agentbla, Ahutson, Ajaxspray, Alanraywiki, Alansohn, AlexiusHoratius, Alfion, AllStarZ, AllanDevil92, Altermike, Ambrose1852, And003, AniRaptor2001, Animum, ApolloBoy, Archtransit, ArielGold, Ark25, Arpingstone, Artrix, Aude, Auroranorth, Auto101, B64, BSI, Bal00, Bavaria, Bavaria II, Beefyt, Beland, Bfoaz, BilCat, Bjenks, Bkissin, Bkonrad, BlackTerror, Blackplate, BlankVerse, BlckKnght, Blubberboy92, Bobblewik, Bobbythe1, Bonadea, BostonRed, BoyoJonesJr, Branddobbe, BrendelSignature, Brithackemack, Brossow, Btornado, Bull-Doser, Bungofpot, Burgundavia, Buron444, BuzzDog, C1010, CZmarlin, Calvin08, CameronTGD, Can't sleep, clown will eat me, CanadianLinuxUser, Capricorn42, Captain Caveman, Car guy 777, Carlb, Carpictures, CaseyPenk, Cdc, Celerityfm, Celosia, Centrx, Cflm001, Cgbolt, Charles01, ChicosBailBonds, Cholmes75, Chowbok, Chris 73, Christopher Mahan, ChristopherF, Chrycoman, Chryslercom, Chryslerforever1988, Comet1440, CommonsDelinker, Conversion script, Cple4, CrazyChemGuy, Crazycomputers, Curps, Cvieg, DBCaravan, DJBMurie, Dan100, Daniela591, Deicchel, DeLarge, Deb, Debresser, Delirium, DerRichter, Desmay, Dhartung, DiRF, Discospinster, DocWatson42, Dodgeram09, Donarreiskoffer, DougW, Dougdodge, Dr Dec, Dr. Blofeld, Dr. zedy, Drewman2011, Drinkybird, Drussel3, Dugwiki, Dynaflow, Dysprosia, Ed Fitzgerald, Ed281685, Edgar181, Edward, Either way, Ejfetters, Ekulwyo, Eliyak, Emerson7, Eog1916, Esrever, EyeSerene, Facts707, FayssalF, Femto, Fitzpatrickjm, Fivestrokes, Fjmustak, Flyer22, Folksong, Frehley, Fuhghettaboutit, GCarty, GDibyendu, GTHO, Gabbe, Georgearison, Gkklein, GoldDragon, Golfj21, Goodnightmush, Gralo, Grammarmonger, Greenshed, Grundle2600, Grunt, Gzuckier, HENNIG1, Hateless, Hbdragon88, Hcobb, Heegoop, Helmandsare, Hmains, Hooperbloob, Hotlorp, Hu12, Hut 8.5, Huw Powell, HybridBoy, Hydrogen Iodide, Iain100, Ianris, Icsunonove, Iggi07, Ikip, Imaginary heroes, Imasleepviking, Infrogmation, Inspiradventure, Ipigott, Isaac Rabinovitch, Isnow, Italianboy10, J, J.delanoy, J3ff, JCDenton2052, JNW, JackLumber, Jacob Poon, Jacobw, JaimeyWB, Jamcib, Janus657, Jcast11, Jeffreymcmanus, Jgera5, Jimp, Jlange, Jmcd88, Jnelson09, Joancreus, JoanneB, John, Jojhutton, Jok2000, Jonrev, Joolz, Jordancpeterson, JosefBranson, Joseph Solis in Australia, Jossi, JuWiki2, Julesd, Jusdafax, Just64helpin, KAMiKAZOW, KansasCity, Kariteh, Karrmann, Kbdank71, Ken6en, Kennedy & Co., Ketiltrout, Kgasso, Kingofdawild166, Kozuch, Krsstowne, Kurykh, Kyuko, Kznf, LSX, Lazylaces, Levineps, Lglswe, Liekersnipes, Lightmouse, Lionel.Mandrake, Llamasharmafarmerdrama, LorenzoB, Lotus L-12, Lowe4091, LtNOWIS, Lucky 6.9, MHSINDIANS, MR42HH, Mac, Magnus Manske, Malcolma, Malepheasant, Mark83, Marshall Stax, Martarius, Martellko, Master Jay, Mathman08, MattKeegan, Mattbr, Mav, Mayccc3, Mboverload, Mbutts, Mdotley, MegX, Mesto, Miami33139, Midwesttoolguy, Mike Rosoft, Millerjoel, Millionairex, Mimihitam, Mindmatrix, Minesweeper, Mintleaf, Miquonranger03, MisfitToys, Mmoneypenny, Morven, MrDolomite, Mrclassicauto, Mtaylor848, Munboy, Mycroft.Holmes, Mysdaao, N2e, N328KF, NCurse, NHRHS2010, NaBUru38, Nafango2, Navstar, NawlinWiki, NcSchu, New England, Niagara, Nicko90161, Nlu, Nopetro, Novous, Nterprize, Oceangor1, OfficerPhil, Oilpanhands, Oknazevad, Old Moonraker, Onetwo1, Ottava Rima, Outriggr, OwenX, PEMC, PLawrence99cx, Paarth1994, Passportguy, Pc13, Philip Trueman, Phonicsmonkey3, Pil56, Pjbflynn, Poppy.langford, Porqin, Progolfplaya247, ProhibitOnions, Quidam65, RadicalBender, RainbowOfLight, Rairdon, Random-Hero-, RandomHero129, Rcandelori, Redd Dragon, Redvers, Reedy, Reflex Reaction, Regushee, Rehrenberg, RevelationDirect, Reverendlinux, Richard1990, Richardshusr, RingtailedFox, RivGuySC, Rjd0060, Rjecina, Rjwilmsi, Rlandmann, Rmsoule, Robert K S, Roentgenium111, Ronald20, Ronhjones, Ronnotel, Roverfan77, Rupertslander, Ryangoff, SB Johnny, SVTCobra, Salamurai, Sameera, Sandahl, Sannse, Scheinwerfermann, SchfiftyThree, SchuminWeb, Scooby1316, Security Analyst, Sertrel, Sfoskett, Sherool, Shortride, Sibbsy, SimonX, Sionus, Skierpage, Sligocki, Slo-mo, Smash, Smashville, Soakologist, Sohailstyle, Sonett72, Speer320, Spinnut, Squash Racket, SriMesh, Staffelde, StaticGull, Steveozone, Stude62, Suldrew, SunKing, Susvolans, Swamilive, Swaq, Sylvain1972, Szyslak, T-dot, THEunique, Tabletop, Take Me Higher, Tanner Kosowan, Tartanperil, Tcotrel, TeaDrinker, TeemPlayer, Teeninvestor, Texaswebscout, The Devil's Advocate, The Flying Spaghetti Monster, The Helper S, The Thing That Should Not Be, Thedjatclubrock, Themanwithoutapast, Theresa knott, Thesmokingmonkey, ThinkEnemies, Thomas Paine1776, Three-quarter-ten, Time, Tingrin87, Tkgd2007, Tohd8BohaithuGh1, Tomtom9041, Tracer9999, Txtrooper, Tydwil3, Typ932, Uberprufen, Ukexpat, Una Smith, Urbanrenewal, Urmomsquared, VK35, Vchimpanzee, Ve2jgs, Vegaswikian, Viddea9, ViperSnake151, Viva Chile, Vossanova, Warren, Watemark, WhatamIdoing, WhisperToMe, Wiarthurhu, WikHead, WikiDon, Wikignome0529, Wikited, Wildtornado, Willgee, Wog7777, Woohookitty, World, Worldkacitizen, Wuhwuzdat, Xavierorr, Xnatedawgx, Xxwikierxx, Yegorm, Yellowdesk, Yizhenwilliam, Yobusdis, Zeamays, 717 anonymous edits

Muscle car *Source*: http://en.wikipedia.org/w/index.php?title=Muscle_car *Contributors*: 123omgnoob123, 293.xx.xxx.xx, A little insignificant, ABVS1936, Aabbcc334455, Ac101, Advds, Ahoerstemeier, Aj510, Alansohn, AmishSexy, AngelofMusic07, Anger22, ArgentLA, Ariedartin, ArmadilloFromHell, ArmyDuck, Asterion, Atlpat, Ayocee, Beetstra, Belovedfreak, Bloodshedder, Blue520, Bmx442, Bobblewik, Bobo192, Bolt Vanderhuge, BorgHunter, Brak44, BrendelSignature, Brossow, Brothejr, C2X, CJ DUB, CZmarlin, Can't sleep, clown will eat me, Canderson7, Carnage99, Casmith 789, Chris Henniker, Christopherlin, Civil Engineer III, Closedmouth, CommonsDelinker, Communitynations, ConradPino, Corvus309, Craigy144, CristianRaiber, Czj, D.brodale, Daniel J. Leivick, Dar-Ape, Darklilac, David R. Ingham, Davy boy1972, Daymark, Deathparade32, DerHexer, DesignExplosion, Discospinster, DocWatson42, Doctorevil64, Dread Specter, Dysepsion, E55Charger, EagleFan, Editor2020, Either way, EurekaLott, Feezo, Fish and karate, Frank-n-stein, Freesek, Friday, GRAHAMUK, Gaius Cornelius, Geneb1955, Gintautasm, Gmail456, Gogo Dodo, GraemeL, Graibeard, Groggy Dice, Gurch, Gus Polly, Hairy Dude, Hezery99, Hiyawathadan, Hmnuser, Hooperbloob, Hu12, Interiot, Irongoxi, Ispy1981, J.delanoy, Jack Tipping, Jamoo, Jaysprout, Jbrian80, Jdmdreamz94, Jgp, Joel Russ, John Nevard, Johnmcenroy, Jusdafax, Jwess, JzG, Karrmann, Keilana, Kevyn, Klichka, Kslyons, Ktr101, Latka, Lavenderbunny, LedgendGamer, Liftarn, Lph, Lunkwill, Lwalt, Mac, MattieTK, Mcs396, Meekywiki, Mephistau, Metroplex, Miex06, Moochie repperton, Moparman68, Morven, Mosn1, Mpotter, MrOllie, MuzMister, Mwanner, Nakon, NateM4488, Nemo, Nhesketh, Nickersonl, Nicsye, Nosaze, Nsaa, Nuttycoconut, O.o, OCDPard, Oilpanhands, Patstuart, Paul foord, Philip Trueman, Plasticup, Porqin, Porsche997SBS, Racer741, Rama, Randroide, Rbwik, Riana, RivGuySC, Rjg6cb, Rjwilmsi, Rnkibbe, Robert Merkel, Robertkeller, Runner5k, SGJ91, Sable232, Sam Korn, Scottmandrake, Sfoskett, Shogunpk, SimonX, Sintaku, Snowwolfslog1, Stephenb, Subash.chandran007, Supernova 6969, Svennex, Swid, TBustah, Tannin, TastyPoutine, Tesseran, That-Vela-Fella, The Editor General, The Thing That Should Not Be, Thingg, Time, Tipdrip215, Tommythegun, Toryt0821, Tresiden, Tronno, Truthseeker7, Typ932, Usmc835, Vegaswikian, Vics3227, Vkt183, Vsmith, Waggers, Warnester, Welsh, WereSpielChequers, Wi-king, Wiarthurhu, Wikiuser100, Windymilla, Wmahan, Writegeist, Xanzzibar, Xbgs351, Xen

Article Sources and Contributors

1986, Xerog, Zach4636, Zchris87v, 627 anonymous edits

List of automobile model nameplates with a discontiguous timeline *Source*: http://en.wikipedia.org/w/index.php?title=List_of_automobile_model_nameplates_with_a_discontiguous_timeline *Contributors*: Buf7579, Donnie Park, Ejjski, Escape Orbit, GregorB, Karrmann, Magasin, Mephistau, Newportryan, Robofish, Szyslak, Taylorr, Weetbixkid, Willirennen, 19 anonymous edits

Image Sources, Licenses and Contributors

Image:Dodge.383.magnum-black.front.view-sstvwf.JPG *Source*: http://en.wikipedia.org/w/index.php?title=File:Dodge.383.magnum-black.front.view-sstvwf.JPG *License*: GNU General Public License *Contributors*: Infrogmation, Owly K, Ranger, Writegeist, 1 anonymous edits

Image:1965-Dodge-Charger-II-Rear.jpg *Source*: http://en.wikipedia.org/w/index.php?title=File:1965-Dodge-Charger-II-Rear.jpg *License*: unknown *Contributors*: Bill Wrigley, CJLL Wright, FastbackJon, Johnmc, Koavf, Manway, Nv8200p, Peripitus, Sable232, Shanes, Undead warrior, 2 anonymous edits

Image:'67 Dodge Charger (Orange Julep).JPG *Source*: http://en.wikipedia.org/w/index.php?title=File:'67_Dodge_Charger_(Orange_Julep).JPG *License*: Public Domain *Contributors*: User:Bull-Doser

Image:66ChargerDash.jpg *Source*: http://en.wikipedia.org/w/index.php?title=File:66ChargerDash.jpg *License*: unknown *Contributors*: Gamgee, Inductiveload, Panzerfaust, Pieter Kuiper

Image:66ChargerSpoiler.JPG *Source*: http://en.wikipedia.org/w/index.php?title=File:66ChargerSpoiler.JPG *License*: Public Domain *Contributors*: Original uploader was FastbackJon at en.wikipedia

Image:Dodge Charger RT 1968.Front.jpg *Source*: http://en.wikipedia.org/w/index.php?title=File:Dodge_Charger_RT_1968.Front.jpg *License*: unknown *Contributors*: Hannes Drexl

Image:Dodge-Charger-1969-Front.jpg *Source*: http://en.wikipedia.org/w/index.php?title=File:Dodge-Charger-1969-Front.jpg *License*: Creative Commons Attribution-Sharealike 2.5 *Contributors*: Aman, Dha, Infrogmation, Liftarn, 3 anonymous edits

Image:Charg.jpg *Source*: http://en.wikipedia.org/w/index.php?title=File:Charg.jpg *License*: unknown *Contributors*: Liftarn, Moto100

File:Dodge Charger 500 (Orange Julep).jpg *Source*: http://en.wikipedia.org/w/index.php?title=File:Dodge_Charger_500_(Orange_Julep).jpg *License*: Public Domain *Contributors*: Bull-Doser

Image:Dodge Charger R-T (Orange Julep).JPG *Source*: http://en.wikipedia.org/w/index.php?title=File:Dodge_Charger_R-T_(Orange_Julep).JPG *License*: Public Domain *Contributors*: User:Bull-Doser

Image:Dodge Charger (Orange Julep).JPG *Source*: http://en.wikipedia.org/w/index.php?title=File:Dodge_Charger_(Orange_Julep).JPG *License*: Public Domain *Contributors*: User:Bull-Doser

Image:Dodge Charger RT 1968.Heck.jpg *Source*: http://en.wikipedia.org/w/index.php?title=File:Dodge_Charger_RT_1968.Heck.jpg *License*: unknown *Contributors*: Hannes Drexl

Image:1973 Dodge Charger SE opera windows.jpg *Source*: http://en.wikipedia.org/w/index.php?title=File:1973_Dodge_Charger_SE_opera_windows.jpg *License*: Creative Commons Attribution-Sharealike 2.0 *Contributors*: User:Howcheng, User:Morven

Image:1969 Dodge Charger green F.jpg *Source*: http://en.wikipedia.org/w/index.php?title=File:1969_Dodge_Charger_green_F.jpg *License*: Public Domain *Contributors*: User:CZmarlin

Image:Benz-velo.jpg *Source*: http://en.wikipedia.org/w/index.php?title=File:Benz-velo.jpg *License*: GNU Free Documentation License *Contributors*: User:Softeis

Image:2000cardistribution.svg *Source*: http://en.wikipedia.org/w/index.php?title=File:2000cardistribution.svg *License*: Public Domain *Contributors*: User:Nevetsjc

Image:World vehicles per capita.svg *Source*: http://en.wikipedia.org/w/index.php?title=File:World_vehicles_per_capita.svg *License*: Public Domain *Contributors*: User:TastyCakes

Image:CarlBenz.jpg *Source*: http://en.wikipedia.org/w/index.php?title=File:CarlBenz.jpg *License*: Public Domain *Contributors*: -

Image:1885Benz.jpg *Source*: http://en.wikipedia.org/w/index.php?title=File:1885Benz.jpg *License*: Public Domain *Contributors*: Infrogmation, Jaranda, Milkmandan, Saforrest, Semnoz, Taisyo

Image:Olds2.jpg *Source*: http://en.wikipedia.org/w/index.php?title=File:Olds2.jpg *License*: Public Domain *Contributors*: Original uploader was Karrmann at en.wikipedia

Image:Henry Ford.jpg *Source*: http://en.wikipedia.org/w/index.php?title=File:Henry_Ford.jpg *License*: Public Domain *Contributors*: CREDIT: Hartsook, photographer. "[Henry Ford, head-and-shoulders portrait, facing slightly left] / Hartsook photo." 1919(?). Prints and Photographs Division, Library of Congress

Image:Late model Ford Model T.jpg *Source*: http://en.wikipedia.org/w/index.php?title=File:Late_model_Ford_Model_T.jpg *License*: GNU Free Documentation License *Contributors*: A. Balet, Editor at Large, Infrogmation, Makary, Mariordo, Nagy, Para, Parkerdr, Werewombat, 8 anonymous edits

File:CNG propelled radio taxi.jpg *Source*: http://en.wikipedia.org/w/index.php?title=File:CNG_propelled_radio_taxi.jpg *License*: Public Domain *Contributors*: Original uploader was Arpanjolly at en.wikipedia

Image:07-Mini-Cooper.jpg *Source*: http://en.wikipedia.org/w/index.php?title=File:07-Mini-Cooper.jpg *License*: Public Domain *Contributors*: IFCAR

Image:TOYOTA FCHV 01.jpg *Source*: http://en.wikipedia.org/w/index.php?title=File:TOYOTA_FCHV_01.jpg *License*: GNU Free Documentation License *Contributors*: User:Ginsin

Image:Kilowatt.jpg *Source*: http://en.wikipedia.org/w/index.php?title=File:Kilowatt.jpg *License*: Public Domain *Contributors*: User DRoberson on en.wikipedia

Image:TeslaRoadster-front.jpg *Source*: http://en.wikipedia.org/w/index.php?title=File:TeslaRoadster-front.jpg *License*: Creative Commons Attribution-Sharealike 2.0 *Contributors*: fogcat5

Image:Catvertroquette.jpg *Source*: http://en.wikipedia.org/w/index.php?title=File:Catvertroquette.jpg *License*: Public Domain *Contributors*: User:Deepak

Image:ABQ RIDE 332 Montgomery Albuquerque.jpg *Source*: http://en.wikipedia.org/w/index.php?title=File:ABQ_RIDE_332_Montgomery_Albuquerque.jpg *License*: GNU Free Documentation License *Contributors*: Original uploader was Camerafiend at en.wikipedia

Image:Car crash 2.jpg *Source*: http://en.wikipedia.org/w/index.php?title=File:Car_crash_2.jpg *License*: Public Domain *Contributors*: Jaranda, MB-one, Thuc, 1 anonymous edits

Image:Dodge logo.svg *Source*: http://en.wikipedia.org/w/index.php?title=File:Dodge_logo.svg *License*: logo *Contributors*: Malcolma, Opertinicy

Image:1915-dodge-archives.jpg *Source*: http://en.wikipedia.org/w/index.php?title=File:1915-dodge-archives.jpg *License*: Attribution *Contributors*: Original uploader was DougW at en.wikipedia

Image:RoyalDodge.jpg *Source*: http://en.wikipedia.org/w/index.php?title=File:RoyalDodge.jpg *License*: Public Domain *Contributors*: TexasAndroid, Tillman

Image:Dodge Series 124 4-Door Sedan 1927.jpg *Source*: http://en.wikipedia.org/w/index.php?title=File:Dodge_Series_124_4-Door_Sedan_1927.jpg *License*: unknown *Contributors*: Lars-Göran Lindgren Sweden

Image:Dodge D11 Luxury Liner 4-Door Sedan 1939.jpg *Source*: http://en.wikipedia.org/w/index.php?title=File:Dodge_D11_Luxury_Liner_4-Door_Sedan_1939.jpg *License*: unknown *Contributors*: Lars-Göran Lindgren Sweden

Image:Dodge Coronet 1955.jpg *Source*: http://en.wikipedia.org/w/index.php?title=File:Dodge_Coronet_1955.jpg *License*: unknown *Contributors*: Lars-Göran Lindgren Sweden

Image:Dodge Coronet 1958.jpg *Source*: http://en.wikipedia.org/w/index.php?title=File:Dodge_Coronet_1958.jpg *License*: unknown *Contributors*: Lglsww

Image:1966 Dodge Coronet.jpg *Source*: http://en.wikipedia.org/w/index.php?title=File:1966_Dodge_Coronet.jpg *License*: Public Domain *Contributors*: User:Sfoskett

Image:1977Diplomat.jpg *Source*: http://en.wikipedia.org/w/index.php?title=File:1977Diplomat.jpg *License*: Public Domain *Contributors*: 328cia, ApolloBoy, MB-one, Thomas doerfer

Image:Dodge Aries sedan.jpg *Source*: http://en.wikipedia.org/w/index.php?title=File:Dodge_Aries_sedan.jpg *License*: Public Domain *Contributors*: IFCAR

Image:Spirit RT Front34.jpg *Source*: http://en.wikipedia.org/w/index.php?title=File:Spirit_RT_Front34.jpg *License*: Public Domain *Contributors*: Original uploader was Scheinwerfermann at en.wikipedia

File:1996 Dodge Stratus ES.JPG *Source*: http://en.wikipedia.org/w/index.php?title=File:1996_Dodge_Stratus_ES.JPG *License*: unknown *Contributors*: User:GreatestWisdom

Image:Black Charger SRT.JPG *Source*: http://en.wikipedia.org/w/index.php?title=File:Black_Charger_SRT.JPG *License*: Public Domain *Contributors*: User MikeElliott01 on en.wikipedia

Image:89RamF34.jpg *Source*: http://en.wikipedia.org/w/index.php?title=File:89RamF34.jpg *License*: Public Domain *Contributors*: Original uploader was Scheinwerfermann at en.wikipedia

Image:DodgeBros.JPG *Source*: http://en.wikipedia.org/w/index.php?title=File:DodgeBros.JPG *License*: Public Domain *Contributors*: Dyno Tested

Image:DSCN6089.JPG *Source*: http://en.wikipedia.org/w/index.php?title=File:DSCN6089.JPG *License*: unknown *Contributors*: Jtrevan, Quadell

Image:ForwardLookPatch.jpg *Source*: http://en.wikipedia.org/w/index.php?title=File:ForwardLookPatch.jpg *License*: unknown *Contributors*: Hammersoft, Scheinwerfermann

Image:Fratzog.jpg *Source*: http://en.wikipedia.org/w/index.php?title=File:Fratzog.jpg *License*: unknown *Contributors*: Stude62, 1 anonymous edits

Image:Dodge Red Pentastar.jpg *Source*: http://en.wikipedia.org/w/index.php?title=File:Dodge_Red_Pentastar.jpg *License*: Public Domain *Contributors*: User:Scheinwerfermann

Image:1978-1979 Chrysler Cordoba.jpg *Source*: http://en.wikipedia.org/w/index.php?title=File:1978-1979_Chrysler_Cordoba.jpg *License*: Public Domain *Contributors*: User:IFCAR

Image:Chrysler Cordoba blue front.jpg *Source*: http://en.wikipedia.org/w/index.php?title=File:Chrysler_Cordoba_blue_front.jpg *License*: Public Domain *Contributors*: User:CZmarlin

Image:ChryslerCordoba.jpg *Source*: http://en.wikipedia.org/w/index.php?title=File:ChryslerCordoba.jpg *License*: GNU Free Documentation License *Contributors*: Sam Krieg

File:'80-'81 Chrysler Cordoba LS (Orange Julep).jpg *Source*: http://en.wikipedia.org/w/index.php?title=File:'80-'81_Chrysler_Cordoba_LS_(Orange_Julep).jpg *License*: Public Domain *Contributors*: Bull-Doser

Image Sources, Licenses and Contributors

Image:Chrysler LLC logo.svg *Source*: http://en.wikipedia.org/w/index.php?title=File:Chrysler_LLC_logo.svg *License*: logo *Contributors*: Blubberboy92

File:Chrysler Headquarters Auburn Hills 20060624.jpg *Source*: http://en.wikipedia.org/w/index.php?title=File:Chrysler_Headquarters_Auburn_Hills_20060624.jpg *License*: Public Domain *Contributors*: Johannes Fasolt

Image:Pontiac GTO 1966.jpg *Source*: http://en.wikipedia.org/w/index.php?title=File:Pontiac_GTO_1966.jpg *License*: unknown *Contributors*: w:User:FreesekFreesek from En-Wikipedia

Image:Plymouth Roadrunner.jpg *Source*: http://en.wikipedia.org/w/index.php?title=File:Plymouth_Roadrunner.jpg *License*: GNU Free Documentation License *Contributors*: User Morven on en.wikipedia

Image:Rocket v8.jpg *Source*: http://en.wikipedia.org/w/index.php?title=File:Rocket_v8.jpg *License*: Public Domain *Contributors*: Original uploader was Zandome at en.wikipedia

Image:Hudson Hornet Club Coupe 1951.jpg *Source*: http://en.wikipedia.org/w/index.php?title=File:Hudson_Hornet_Club_Coupe_1951.jpg *License*: unknown *Contributors*: Lglswe, Morio, Writegeist

Image:Chrysler C-300.jpg *Source*: http://en.wikipedia.org/w/index.php?title=File:Chrysler_C-300.jpg *License*: GNU Free Documentation License *Contributors*: Original uploader was w:en:User:MorvenMorven at en.wikipedia

Image:1957 Rambler Rebel front.JPG *Source*: http://en.wikipedia.org/w/index.php?title=File:1957_Rambler_Rebel_front.JPG *License*: Public Domain *Contributors*: User:CZmarlin

Image:A 1964 Ford Thunderbolt Muscle Car.jpg *Source*: http://en.wikipedia.org/w/index.php?title=File:A_1964_Ford_Thunderbolt_Muscle_Car.jpg *License*: unknown *Contributors*: User:Writegeist

Image:1970redGTX.JPG *Source*: http://en.wikipedia.org/w/index.php?title=File:1970redGTX.JPG *License*: GNU Free Documentation License *Contributors*: User Sirboxxer on en.wikipedia

Image:1970 AMC Rebel Machine Muscle Car-RWB.jpg *Source*: http://en.wikipedia.org/w/index.php?title=File:1970_AMC_Rebel_Machine_Muscle_Car-RWB.jpg *License*: Public Domain *Contributors*: Christopher Ziemnowicz

Image:1970-1971 Holden HG Monaro GTS 01.jpg *Source*: http://en.wikipedia.org/w/index.php?title=File:1970-1971_Holden_HG_Monaro_GTS_01.jpg *License*: unknown *Contributors*: -

Image:Ford Cobra no.66.jpg *Source*: http://en.wikipedia.org/w/index.php?title=File:Ford_Cobra_no.66.jpg *License*: unknown *Contributors*: Original uploader was Cobra066 at en.wikipedia

Image:Ford Capri (1974) Cosworth.jpg *Source*: http://en.wikipedia.org/w/index.php?title=File:Ford_Capri_(1974)_Cosworth.jpg *License*: Creative Commons Attribution-Sharealike 2.0 *Contributors*: User:LSDSL

Image:94-96 Chevrolet Impala SS.jpg *Source*: http://en.wikipedia.org/w/index.php?title=File:94-96_Chevrolet_Impala_SS.jpg *License*: Public Domain *Contributors*: IFCAR

Image:Mercury Marauder.jpg *Source*: http://en.wikipedia.org/w/index.php?title=File:Mercury_Marauder.jpg *License*: Public Domain *Contributors*: IFCAR

VDM publishing house ltd.

Scientific Publishing House
offers
free of charge publication

of current academic research papers, Bachelor´s Theses, Master's Theses, Dissertations or Scientific Monographs

If you have written a thesis which satisfies high content as well as formal demands, and you are interested in a remunerated publication of your work, please send an e-mail with some initial information about yourself and your work to *info@vdm-publishing-house.com*.

Our editorial office will get in touch with you shortly.

VDM Publishing House Ltd.
Meldrum Court 17.
Beau Bassin
Mauritius
www.vdm-publishing-house.com

GNU Free Documentation License Version 1.2, November 2002 Copyright (C) 2000,2001,2002 Free Software Foundation, Inc. 59 Temple Place, Suite 330, Boston, MA 02111-1307 USA Everyone is permitted to copy and distribute verbatim copies of this license document, but changing it is not allowed.

0. PREAMBLE

The purpose of this License is to make a manual, textbook, or other functional and useful document "free" in the sense of freedom: to assure everyone the effective freedom to copy and redistribute it, with or without modifying it, either commercially or noncommercially. Secondarily, this License preserves for the author and publisher a way to get credit for their work, while not being considered responsible for modifications made by others. This License is a kind of "copyleft", which means that derivative works of the document must themselves be free in the same sense. It complements the GNU General Public License, which is a copyleft license designed for free software. We have designed this License in order to use it for manuals for free software, because free software needs free documentation: a free program should come with manuals providing the same freedoms that the software does. But this License is not limited to software manuals; it can be used for any textual work, regardless of subject matter or whether it is published as a printed book. We recommend this License principally for works whose purpose is instruction or reference.

1. APPLICABILITY AND DEFINITIONS

This License applies to any manual or other work, in any medium, that contains a notice placed by the copyright holder saying it can be distributed under the terms of this License. Such a notice grants a world-wide, royalty-free license, unlimited in duration, to use that work under the conditions stated herein. The "Document", below, refers to any such manual or work. Any member of the public is a licensee, and is addressed as "you". You accept the license if you copy, modify or distribute the work in a way requiring permission under copyright law. A "Modified Version" of the Document means any work containing the Document or a portion of it, either copied verbatim, or with modifications and/or translated into another language. A "Secondary Section" is a named appendix or a front-matter section of the Document that deals exclusively with the relationship of the publishers or authors of the Document to the Document's overall subject (or to related matters) and contains nothing that could fall directly within that overall subject. (Thus, if the Document is in part a textbook of mathematics, a Secondary Section may not explain any mathematics.) The relationship could be a matter of historical connection with the subject or with related matters, or of legal, commercial, philosophical, ethical or political position regarding them. The "Invariant Sections" are certain Secondary Sections whose titles are designated, as being those of Invariant Sections, in the notice that says that the Document is released under this License. If a section does not fit the above definition of Secondary then it is not allowed to be designated as Invariant. The Document may contain zero Invariant Sections. If the Document does not identify any Invariant Sections then there are none. The "Cover Texts" are certain short passages of text that are listed, as Front-Cover Texts or Back-Cover Texts, in the notice that says that the Document is released under this License. A Front-Cover Text may be at most 5 words, and a Back-Cover Text may be at most 25 words. A "Transparent" copy of the Document means a machine-readable copy, represented in a format whose specification is available to the general public, that is suitable for revising the document straightforwardly with generic text editors or (for images composed of pixels) generic paint programs or (for drawings) some widely available drawing editor, and that is suitable for input to text formatters or for automatic translation to a variety of formats suitable for input to text formatters. A copy made in an otherwise Transparent file format whose markup, or absence of markup, has been arranged to thwart or discourage subsequent modification by readers is not Transparent. An image format is not Transparent if used for any substantial amount of text. A copy that is not "Transparent" is called "Opaque". Examples of suitable formats for Transparent copies include plain ASCII without markup, Texinfo input format, LaTeX input format, SGML or XML using a publicly available DTD, and standard-conforming simple HTML, PostScript or PDF designed for human modification. Examples of transparent image formats include PNG, XCF and JPG. Opaque formats include proprietary formats that can be read and edited only by proprietary word processors, SGML or XML for which the DTD and/or processing tools are not generally available, and the machine-generated HTML, PostScript or PDF produced by some word processors for output purposes only. The "Title Page" means, for a printed book, the title page itself, plus such following pages as are needed to hold, legibly, the material this License requires to appear in the title page. For works in formats which do not have any title page as such, "Title Page" means the text near the most prominent appearance of the work's title, preceding the beginning of the body of the text. A section "Entitled XYZ" means a named subunit of the Document whose title either is precisely XYZ or contains XYZ in parentheses following text that translates XYZ in another language. (Here XYZ stands for a specific section name mentioned below, such as "Acknowledgements", "Dedications", "Endorsements", or "History".) To "Preserve the Title" of such a section when you modify the Document means that it remains a section "Entitled XYZ" according to this definition. The Document may include Warranty Disclaimers next to the notice which states that this License applies to the Document. These Warranty Disclaimers are considered to be included by reference in this License, but only as regards disclaiming warranties: any other implication that these Warranty Disclaimers may have is void and has no effect on the meaning of this License.

2. VERBATIM COPYING

You may copy and distribute the Document in any medium, either commercially or noncommercially, provided that this License, the copyright notices, and the license notice saying this License applies to the Document are reproduced in all copies, and that you add no other conditions whatsoever to those of this License. You may not use technical measures to obstruct or control the reading or further copying of the copies you make or distribute. However, you may accept compensation in exchange for copies. If you distribute a large enough number of copies you must also follow the conditions in section 3. You may also lend copies, under the same conditions stated above, and you may publicly display copies.

3. COPYING IN QUANTITY

If you publish printed copies (or copies in media that commonly have printed covers) of the Document, numbering more than 100, and the Document's license notice requires Cover Texts, you must enclose the copies in covers that carry, clearly and legibly, all these Cover Texts: Front-Cover Texts on the front cover, and Back-Cover Texts on the back cover. Both covers must also clearly and legibly identify you as the publisher of these copies. The front cover must present the full title with all words of the title equally prominent and visible. You may add other material on the covers in addition. Copying with changes limited to the covers, as long as they preserve the title of the Document and satisfy these conditions, can be treated as verbatim copying in other respects. If the required texts for either cover are too voluminous to fit legibly, you should put the first ones listed (as many as fit reasonably) on the actual cover, and continue the rest onto adjacent pages. If you publish or distribute Opaque copies of the Document numbering more than 100, you must either include a machine-readable Transparent copy along with each Opaque copy, or state in or with each Opaque copy a computer-network location from which the general network-using public has access to download using public-standard network protocols a complete Transparent copy of the Document, free of added material. If you use the latter option, you must take reasonably prudent steps, when you begin distribution of Opaque copies in quantity, to ensure that this Transparent copy will remain thus accessible at the stated location until at least one year after the last time you distribute an Opaque copy (directly or through your agents or retailers) of that edition to the public. It is requested, but not required, that you contact the authors of the Document well before redistributing any large number of copies, to give them a chance to provide you with an updated version of the Document.

4. MODIFICATIONS

You may copy and distribute a Modified Version of the Document under the conditions of sections 2 and 3 above, provided that you release the Modified Version under precisely this License, with the Modified Version filling the role of the Document, thus licensing distribution and modification of the Modified Version to whoever possesses a copy of it. In addition, you must do these things in the Modified Version: A. Use in the Title Page (and on the covers, if any) a title distinct from that of the Document, and from those of previous versions (which should, if there were any, be listed in the History section of the Document). You may use the same title as a previous version if the original publisher of that version gives permission. B. List on the Title Page, as authors, one or more persons or entities responsible for authorship of the modifications in the Modified Version, together with at least five of the principal authors of the Document (all of its principal authors, if it has fewer than five), unless they release you from this requirement. C. State on the Title page the name of the publisher of the Modified Version, as the publisher. D. Preserve all the copyright notices of the Document. E. Add an appropriate copyright notice for your modifications adjacent to the other copyright notices. F. Include, immediately after the copyright notices, a license notice giving the public permission to use the Modified Version under the terms of this License, in the form shown in the Addendum below. G. Preserve in that license notice the full lists of Invariant Sections and required Cover Texts given in the Document's license notice. H. Include an unaltered copy of this License. I. Preserve the section Entitled "History", Preserve its Title, and add to it an item stating at least the title, year, new authors, and publisher of the Modified Version as given on the Title Page. If there is no section Entitled "History" in the Document, create one stating the title, year, authors, and publisher of the Document as given on its Title Page, then add an item describing the Modified Version as stated in the previous sentence. J. Preserve the network location, if any, given in the Document for public access to a Transparent copy of the Document, and likewise the network locations given in the Document for previous versions it was based on. These may be placed in the "History" section. You may omit a network location for a work that was published at least four years before the Document itself, or if the original publisher of the version it refers to gives permission. K. For any section Entitled "Acknowledgements" or "Dedications", Preserve the Title of the section, and preserve in the section all the substance and tone of each of the contributor acknowledgements and/or dedications given therein. L. Preserve all the Invariant Sections of the Document, unaltered in their text and in their titles. Section numbers or the equivalent are not considered part of the section titles. M. Delete any section Entitled "Endorsements". Such a section may not be included in the Modified Version. N. Do not retitle any existing section to be Entitled "Endorsements" or to conflict in title with any Invariant Section. O. Preserve any Warranty Disclaimers. If the Modified Version includes new front-matter sections or appendices that qualify as Secondary Sections and contain no material copied from the Document, you may at your option designate some or all of these sections as invariant. To do this, add their titles to the list of Invariant Sections in the Modified Version's license notice. These titles must be distinct from any other section titles. You may add a section Entitled "Endorsements", provided it contains nothing but endorsements of your Modified Version by various parties--for example, statements of peer review or that the text has been approved by an organization as the authoritative definition of a standard. You may add a passage of up to five words as a Front-Cover Text, and a passage of up to 25 words as a Back-Cover Text, to the end of the list of Cover Texts in the Modified Version. Only one passage of Front-Cover Text and one of Back-Cover Text may be added by (or through arrangements made by) any one entity. If the Document already includes a cover text for the same cover, previously added by you or by arrangement made by the same entity you are acting on behalf of, you may not add another; but
you may replace the old one, on explicit permission from
previous publisher that added the old one. The author(s)
publisher(s) of the Document do not by this License
permission to use their names for publicity for or to assert or i
endorsement of any Modified Version.

5. COMBINING DOCUMENTS

You may combine the Document with other documents rele
under this License, under the terms defined in section 4 abov
modified versions, provided that you include in the combinatio
of the Invariant Sections of all of the original docum
unmodified, and list them all as Invariant Sections of
combined work in its license notice, and that you preserve all
Warranty Disclaimers. The combined work need only contain
copy of this License, and multiple identical Invariant Sections
be replaced with a single copy. If there are multiple Inva
Sections with the same name but different contents, make
title of each such section unique by adding at the end of
parentheses, the name of the original author or publisher of
section if known, or else a unique number. Make the s
adjustment to the section titles in the list of Invariant Sectio
the license notice of the combined work. In the combination,
must combine any sections Entitled "History" in the va
original documents, forming one section Entitled "His
likewise combine any sections Entitled "Acknowledgements",
any sections Entitled "Dedications". You must delete all sec
Entitled "Endorsements".

6. COLLECTIONS OF DOCUMENTS

You may make a collection consisting of the Document and
documents released under this License, and replace
individual copies of this License in the various documents w
single copy that is included in the collection, provided that
follow the rules of this License for verbatim copying of each c
documents in all other respects. You may extract a s
document from such a collection, and distribute it indivic
under this License, provided you insert a copy of this License
the extracted document, and follow this License in all
respects regarding verbatim copying of that document.

7. AGGREGATION WITH INDEPENDENT WORKS

A compilation of the Document or its derivatives with
separate and independent documents or works, in or on a vo
of a storage or distribution medium, is called an "aggregate"
copyright resulting from the compilation is not used to limi
legal rights of the compilation's users beyond what the indiv
works permit. When the Document is included in an aggre
this License does not apply to the other works in the aggre
which are not themselves derivative works of the Document.
Cover Text requirement of section 3 is applicable to these c
of the Document, then if the Document is less than one half c
entire aggregate, the Document's Cover Texts may be place
covers that bracket the Document within the aggregate, o
electronic equivalent of covers if the Document is in elec
form. Otherwise they must appear on printed covers that bra
the whole aggregate.

8. TRANSLATION

Translation is considered a kind of modification, so you
distribute translations of the Document under the terms of se
4. Replacing Invariant Sections with translations requires sp
permission from their copyright holders, but you may in
translations of some or all Invariant Sections in addition t
original versions of these Invariant Sections. You may inclu
translation of this License, and all the license notices i
Document, and any Warranty Disclaimers, provided that you
include the original English version of this License and
original versions of those notices and disclaimers. In case
disagreement between the translation and the original versi
this License or a notice or disclaimer, the original versio
prevail. If a section in the Document is Er
"Acknowledgements", "Dedications", or "History", the require
(section 4) to Preserve its Title (section 1) will typically re
changing the actual title.

9. TERMINATION

You may not copy, modify, sublicense, or distribute
Document except as expressly provided for under this Lic
Any other attempt to copy, modify, sublicense or distribut
Document is void, and will automatically terminate your
under this License. However, parties who have received co
or rights, from you under this License will not have their lice
terminated so long as such parties remain in full compliance.

10. FUTURE REVISIONS OF THIS LICENSE

The Free Software Foundation may publish new, re
versions of the GNU Free Documentation License from tin
time. Such new versions will be similar in spirit to the pr
version, but may differ in detail to address new problen
concerns. See http://www.gnu.org/copyleft/. Each version o
License is given a distinguishing version number. If the Docu
specifies that a particular numbered version of this Licens
any later version" applies to it, you have the option of follo
the terms and conditions either of that specified version or o
later version that has been published (not as a draft) by the
Software Foundation. If the Document does not specify a ve
number of this License, you may choose any version
published (not as a draft) by the Free Software Founda
ADDENDUM: How to use this License for your documents T
this License in a document you have written, include a copy
License in the document and put the following copyright
license notices just after the title page: Copyright (c) YEAR Y
NAME. Permission is granted to copy, distribute and/or m
this document under the terms of the GNU Free Document
License, Version 1.2 or any later version published by the
Software Foundation; with no Invariant Sections, no Front-C
Texts, and no Back-Cover Texts. A copy of the license is incl
in the section entitled "GNU Free Documentation License".
have Invariant Sections, Front-Cover Texts and Back-C
Texts, replace the "with...Texts." line with this: with the Inva
Sections being LIST THEIR TITLES, with the Front-Cover
being LIST, and with the Back-Cover Texts being LIST. I
have Invariant Sections without Cover Texts, or some
combination of the three, merge those two alternatives to su
situation. If your document contains nontrivial example
program code, we recommend releasing these example
parallel under your choice of free software license, such a
GNU General Public License, to permit their use in free softw

Lightning Source UK Ltd.
Milton Keynes UK
UKOW02f0836210815

257305UK00001B/56/P

9 786130 293345